# Science and Fiction

## Science and Fiction – A Springer Series

This collection of entertaining and thought-provoking books will appeal equally to science buffs, scientists and science-fiction fans. It was born out of the recognition that scientific discovery and the creation of plausible fictional scenarios are often two sides of the same coin. Each relies on an understanding of the way the world works, coupled with the imaginative ability to invent new or alternative explanations—and even other worlds. Authored by practicing scientists as well as writers of hard science fiction, these books explore and exploit the borderlands between accepted science and its fictional counterpart. Uncovering mutual influences, promoting fruitful interaction, narrating and analyzing fictional scenarios, together they serve as a reaction vessel for inspired new ideas in science, technology, and beyond.

Whether fiction, fact, or forever undecidable: the Springer Series "Science and Fiction" intends to go where no one has gone before!

Its largely non-technical books take several different approaches. Journey with their authors as they

- Indulge in science speculation – describing intriguing, plausible yet unproven ideas;
- Exploit science fiction for educational purposes and as a means of promoting critical thinking;
- Explore the interplay of science and science fiction – throughout the history of the genre and looking ahead;
- Delve into related topics including, but not limited to: science as a creative process, the limits of science, interplay of literature and knowledge;
- Tell fictional short stories built around well-defined scientific ideas, with a supplement summarizing the science underlying the plot.

Readers can look forward to a broad range of topics, as intriguing as they are important. Here just a few by way of illustration:

- Time travel, superluminal travel, wormholes, teleportation
- Extraterrestrial intelligence and alien civilizations
- Artificial intelligence, planetary brains, the universe as a computer, simulated worlds
- Non-anthropocentric viewpoints
- Synthetic biology, genetic engineering, developing nanotechnologies
- Eco/infrastructure/meteorite-impact disaster scenarios
- Future scenarios, transhumanism, posthumanism, intelligence explosion
- Virtual worlds, cyberspace dramas
- Consciousness and mind manipulation

More information about this series at http://www.springer.com/series/11657

Peter Schattschneider

# The EXODUS Incident

## A Scientific Novel

Springer

Peter Schattschneider
Institute of Solid State Physics
TU Wien
Vienna, Austria

ISSN 2197-1188          ISSN 2197-1196    (electronic)
Science and Fiction
ISBN 978-3-030-70018-8          ISBN 978-3-030-70019-5    (eBook)
https://doi.org/10.1007/978-3-030-70019-5

Cover credits: "Atlantis Landscape" by Lucas Giesinger [https://www.flickr.com/peopl/192057596@N04/], licensed under CC BY 2.0 / A derivative from "The White Pocket" by John Fowler [https://www.flickr.com/photos/snowpeak/] and from "Solar Flare" by NASA, both licensed under CC BY 2.0; Portrait used on back cover: © Klaus Ranger Fotografie.

This Springer imprint is published by the registered company Springer Nature Switzerland AG.
The registered company address is: Gewerbestrasse 11, 6330 Cham, Switzerland

# Foreword

Of course, a futuristic novel first of all serves to entertain the reader, but it is far from being its only effect. Whether or not the author does it intentionally, science fiction, by taking a look at a possible future, presents a model of tomorrow's society and technological development. In this respect, I am convinced that authors of this genre have a great deal of responsibility, which they may not always be able to live up to. But they should at least be aware of this. I do not intend to say that the occurrence of the future described in such stories is a sign of quality. For I am convinced that the future is not foreseeable. What I am saying is rather that the tomorrow as a model presented should be coherent and based on the facts of modern science and technology, i.e. it should be possible in principle. As a reader, you may or may not like this literary model of the future, and the emotions generated by such reading may even encourage some to make an active contribution to shaping this concrete future of ours. From numerous letters and personal contacts I have learned that my science fiction books and stories have inspired young people to study science or technology. Influencing a person's life path in this way is the greatest praise I can receive as an author for my work.

Peter Schattschneider came to science fiction as a schoolboy—and to an encounter with a comet which seemed to intervene almost fatefully and guide him like the Three Wise Men. This comet appears in the title of my first book published in 1960: *The Green Comet*. This collection of short stories inspired the adolescent for the science in science fiction, as he once told me. Peter later studied physics and is now known as professor emeritus, affiliated to the Vienna University of Technology.

Only later did I learn that there was a remarkable point of contact in both our lives: I had done a doctorate in electron optics, and curiously enough, decades later Peter chose electron microscopy as a research field. What is even more remarkable is that Walter Glaser, my then very young doctoral supervisor, after his return from the USA became full professor at the very same institute where Peter Schattschneider does research today.

It seems that *The Green Comet* also showed Peter the way to literature. After numerous unsuccessful apprentice pieces he sent me a story manuscript. The year was 1978. Email was science fiction; the text came by mail. I liked it, so I published it in one of my anthologies.

Since then, in addition to many stories, including award-winning ones, he has written an episodic novel entitled *Singularitäten*, published by Suhrkamp in 1984, and more recently, the science fiction novel *Hell Fever*. For the sake of his scientific career, science fiction remained a hobby besides his earnest life as a solid-state physicist at the Vienna University of Technology. He also held guest professorships at the Centre National de la Recherche Scientifique in Paris and in Toulouse. I also remember that he organised workshops on the relationship between science and science fiction, which I find very remarkable.

Now let us turn to his most recent novel, the one before us. Of course, I will not reveal anything about the content, just a few thoughts that came to my mind. The story deals with issues surrounding the limits of human cognition. It also highlights the extent to which science has long since decoupled itself from the reality in which we live every day. The reader may also have thoughts that revolve around the complexity of twenty-first-century science. To what extent do computer representations of processes in nature—a popular understanding tool in science today—actually have the meaning that some people ascribe to them: as a factual representation of reality?

Schattschneider stimulates such *Gedanken* experiments first and foremost with the fictitious scientific-technical appendix, which contains many highly interesting facets on relativistic space travel, Lorentz contraction, planetary physics and also on the error-proneness of simulations. In the interest of the exciting flow of action, the outsourcing of such technical details was the method of choice.

To conclude, the reader is led into a world that appears to be highly fantastic but at the same time stands firmly on the ground of science and technology—just as I imagine good science fiction to be. I am convinced that this work will find a readership at the prestigious Springer publishing house that not only appreciates the sophisticated entertainment but also motivates the curious one to explore the scientific and technical issues raised in the book in some way or another!

PS: Is it pure coincidence that in the last phrase of the novel the Magi, led by a comet as we know, make a brief appearance again? Or are they even a joke that the programmers of our universe allowed themselves?

Herbert W. Franke

# Preface

To my knowledge, there is no proper definition of science fiction. *SciFi is everything that can be sold as SciFi*, declared Wolfgang Jeschke, my editor at Heyne, and there is little to object to. In the end, this is a banal case of tacit knowledge—every SciFi reader knows, of course, whether he or she has SciFi on the desk or not.

The definition problem becomes more interesting with hard SciFi, i.e. scientifically based fiction. What, exactly, is a hard SciFi story? How can one safely recognise the science? Giant insects, Godzilla or shrinking people are not among them, not because insects cannot be imagined skyscraper-tall and people cannot be imagined miniscule as a microbe, but because the authors usually are not familiar with the laws of scaling. A tarantula enlarged a hundredfold would collapse under the force of gravity because the fracture strength of the legs increases with the square, but the weight increases with the third power of the enlargement factor. That is why evolution quickly abandoned this idea. And so it goes on—right across topics from all kinds of different areas: superluminal speed, teleportation, time travel.... Exciting, amusing and sometimes enlightening, but on closer inspection they lead to phenomena incompatible with the plot, or to paradoxes as for instance in time travel stories, but this is more than often intentional.

Considering the enormous flood of science fiction literature, there are a few works by a few authors that meet the criterion of hard SciFi: Poul Anderson, Isaac Asimov, Gregory Benford, Arthur C. Clarke, Hal Clement, Cixin Liu, Greg Egan, Robert Forward, Gerard Klein, Larry Niven and Jules Verne, to name the most influentials. Often the scenarios are physically well-founded, but only verbally described. The critical reader of Poul Anderson's *Tau zero* might think: "It may well be that *Leonora Christine's* ramjet engine accelerates the space ship with 1 g, but if you don't tell us the density of interstellar hydrogen, the scoop radius, the engine thrust and the ship's mass, you can claim a great deal of nonsense."

That said, only a tiny subset of hard SciFi texts can be checked with math and physics. Let me give just one example: the stellar rainbow, that colourful arc of stars which flits about several works of renowned authors. It is fake news. Certainly, the Doppler effect changes the colour of stars for an observer in a relativistically fast space ship, but the subtle details do not lead to rainbow colours. Why this is so will be revealed to the curious reader in the present novel.

The scientific verifiability of SciFi texts was the starting point for a university lecture with the working title "How Physics inspires SciFi", which I developed in the 1990s in collaboration with colleagues from the École Centrale in Paris. We started with a handful of short stories by H. G. Wells, A. C. Clarke and L. Niven and examined the *physics in fiction* mathematically, with at times surprising results. More and more texts were added over the years, so that the course participants were soon able to check selected stories themselves in small groups with basic physics, intuition and creativity. It was great fun.

I have been writing SciFi since the 1970s. The confidence in technology at that time and the later prevalence of dystopia, not accidentally after the Orwell year 1984, provided a fertile ground, the interest of the public was evident, and I was lucky to find friendly editors such as Herbert W. Franke and Franz Rottensteiner. This resulted in the publication of numerous short stories and two novels. Some of them are hard SciFi, but there is little that can be verified mathematically. This is paradoxical in so far as I used such texts in the lecture which I held for many years. The problem is that my topics mostly deal with epistemology: what is real, where does knowledge come from and how can we rely on it? These questions are not far from discussions about an alarming trend of our days: the lure of conspiracy theories. It is not by chance that the *EXODUS incident* touches this sore spot of our days.

When philosophy meets hard SciFi the author has a problem. After all, what should one calculate in philosophy? Which conservation law applies to epistemology? This problem occupied me for a long time until I had the idea of perfidiously entangling the wealth of math experience from the lecture I had held with my colleagues, all that is verifiable and can be calculated, with my more philosophical topic. Here, entanglement can well be understood in Schrödinger's sense, even if only epistemologically. (Trigger warning: this is a double pun.) Out of it the *EXODUS incident* has come into being. In a way, it was the dialectical resolution of a persistent contradiction. The novel can be read as a detective story in search of truth in the abyss of falsehood, or as a classroom exercise in physics. In fact, the strange phenomena which decorate the plot can be checked with rigid math. The editors at Springer convinced me that so much hardcore physics would go beyond the scope of the series. The interested reader finds the equations and derivations here: https://www.ustem.tuwien.ac.at/exodusincident

During the entire book project I enjoyed support from many sides. I would like to thank my colleagues Pascal Bernaud and Ann-Lenaig Hamon from CentraleSupélec in Paris and Cécile Hébert from the École Polytechnique Fédérale de Lausanne for their careful calculations in reviewing hard science fiction over the years, as well as the students of our lectures in Vienna, Paris and Beijing who critically questioned many texts and gained surprising insights that we ourselves had missed. I would also like to thank Herbert W. Franke—my role model since my first steps as a writer—for his valuable comments, and my agent Franz Rottensteiner who has always given me the best advice. My dear friend Manfred Linke saved me from blathering too much. Lukas Giesinger, good bloke who never says "no" when asked for help, checked the blueprint of the spaceship. His know-how in media design and John Fowler's stunning photograpy made the alien landscape of Atlantis a fantastic experience.

Special thanks go to my test readers Albert Blauensteiner, Herbert "Hörby" Hutter and Brandon Weigel. They discovered a great deal of nonsense in the manuscript. The expertise of Paul Gilster and Al Jackson on the physics of Bussard ramjets was extremely helpful. I am indebted to the team at Springer, especially to Mark Alpert who provided me with excellent support during the phase of translation from German and during proofreading, and to Lisa Scalone for helping me with tedious administrative problems. My thanks are due not least to my partner and all my friends, whom I tormented with tiresome questions and emotional absences during the writing phase. I love you all; you are wonderful!

Vienna, Austria                                         Peter Schattschneider
January, 2021

# Contents

1   The Novel: The EXODUS Incident                        1

2   The EXODUS Incident: A Failure Analysis             157

# 1

# The Novel: The EXODUS Incident

**Abstract** In the near future, Earth is suffering from climate change, famines, and fundamentalism. A global nuclear war is imminent. Interstellar probes from the Breakthrough Starshot project initiated by J. Milner and S. Hawking have discovered a habitable planet in the stellar system Proxima Centauri, just in time for the exodus of the elites. On board the EXODUS starship, the crew starts to experience strange things. The voyage to Atlantis, the new home for mankind, enters a mysterious and disquieting territory, where conspiracy theories about what is real and what is virtual emerge.

*Reality is that which, when you stop believing in it, doesn't go away.*

*Philip K. Dick*

## Episode 1 Special Task Force

#No kidding?
The oil wells run dry, and the Gulf Stream's going to die. Two funny hopes for a climate change.

#AliceWonders
EUROFORCE concentrates troops on the Spanish and Italian borders. People are migrating north. Marauding gangs endanger the borders of core Europe.

Flies buzzed voraciously around the crater. Within 10 meters, the remains of the explosion were scattered in the clearing—tattered clothing, skin fragments, body parts, bone splinters. The body had been ripped apart by a mini-bomb.

Commissioner Oliver Storm wiped the sweat from his forehead and swatted the flies away.

"Damned flies."

"No cursing please!" His colleague Alice Falkenberg insisted. "Speak properly!"

"The flies won't complain."

Disgusted, Storm turned away and checked the area around the bomb crater. They were in a small patch of woodland—gnarled oaks, scattered beech trees, undergrowth, scrub. There, where a tranquil footpath opened to a clearing, lay the victim. The homicide squad's patrol car was visible through the trees and behind it the nearby village: low houses, withered meadows and fields, a church tower.

The forensic robot rolled back to the drone and latched on. The propellers started, the drone took off and disappeared quickly over the woods.

"Do you want to secure anything else?" the Commissioner asked his fellow investigators. The men in protective clothing answered in the negative, closed their suitcases, and set off for the village.

"Okay, Alice, you can send in the street sweepers now."

"Sure thing." She relayed the order to her mobile phone.

Alice poked her foot into the mini-crater that had been torn open by the explosives. Tiny splinters glistening in the sun. Storm bent down, picked up one of the larger ones, and looked at it from all sides.

"Ceramics. Like the others. Same colors, same thickness."

"We'll see. The chemists will tell us exactly."

They circled the site several times hoping to find usable trace evidence. They stopped at the thorax of the torso. The Commissioner pointed at the neck, which was cut clean through. The head was missing.

"Like all the others," murmured Falkenberg. "What's he trying to tell us?"

"Speak properly," Storm mimicked her.

"Huh?"

"How do you know the perpetrator was a man?"

Alice rolled her eyes and said: "All the bodies we've found so far are headless. Is this a signal that his-her victims were too stupid to live?"

"Maybe she's collecting heads. We have, including this one, four female and two male bodies in the series, all decapitated. That's a strong indication of jealousy."

Falkenberg looked at him in astonishment. "Why jealousy?"

"Only if it's a she," he smirked.

"You're so stupid!" She shook her head and trotted to the police car. He followed mechanically.

The analysis of the tracks confirmed what Storm had suspected. Plastic explosives in a ceramic casing had been detonated close to the body. The head was then cut off from the shredded torso. The killer had medical knowledge, as the incisions proved. At the crime scene, neither foot nor tire tracks had been found, only strange indentations like those made by a stick or crutches. Storm pulled the extensive file that Alice had e-mailed him onto the screen. A photo of an attractive woman appeared, including her data.

"Marie Rückert, thirty-two. Saleswoman in a health food store, childless, no ties." Storm nodded. "That's quite something."

"What?"

"Progress. We got this information after only two hours today."

"It's faster when we have a good DNA sample. And it helps that everyone's required to link their personal data to their DNA. When did that become a law anyway?"

"Before your time." *The kids have no idea what it was like before*, he thought. Back when you could go on vacation in Spain. When you could buy a car. Back when things were better.

He wiped the sweat from his brow. "The heat is killing me."

Alice turned the table fan on in his direction. "Patience. The Gulf Stream's going to die. It'll cool down in a couple of years."

"I'll buy you a beer on that." He unbuttoned his shirt collar and leaned into the lukewarm airstream. "Let's go over it again," he suggested weakly.

"We have a saleswoman, a biologist, a nutritionist, an architect, a gunrunner, and an IT expert. The victims are between thirty and forty years old."

"Same age group. The Lost Generation, that's what it was called during the Great Confusion. Not very helpful. The series started with Lorraine Bisset..."

"...the architect. Three years ago, that was. Six months later, Roland Petrides, bachelor's degree in computer science. A week later, Sandra Eckermann, master's degree in biology. Then there's a break—almost a year—until Kelly Clark, the nutritionist, she has a degree from Kings College. Not bad."

Alice continued to scroll through the dossier. He watched her while he turned a pencil between his fingers. Attractive, active, empathic (probably). Everything that he was not. What a handsome couple we'd make—*until you teach me the ins and outs*, he thought in disillusion. At her age, there were two possibilities: career or child. Both would end in pair annihilation. It would be worth it, though, if she was good in bed. A few tricks to get the better of life's dreariness for a short time. No dramatic break-up scenes. When the case was solved, he would leave again. Maybe something could be arranged.

"And six months ago, Otto Freissler," she interrupted his daydream.

"What?"

"Otto Freissler, your old friend."

"The gunrunner. He was actually quite okay. It's a shame it can't be considered an honorable profession."

"All the crime scenes are in the Vienna area. No usable traces except foreign DNA, always the same. Matching the DNA against the sequence database is negative, no match with older hospital or forensic records either. Time of the crime is always at night."

"Witnesses, observations?"

"Nothing useful. Victims met with friends, acquaintances, family, lovers, prostitutes prior to their disappearance. No similarities, no motives, no strong suspicions." Alice shook her head. "The victims disappeared just before the murders. We found abandoned cars on country roads and in parking garages, backpacks on hiking trails, clothes in gardens. The video surveillance analysis was inconclusive."

"What about that case a year ago, what was it called? You know, that upper-class hotel—"

Storm scrolled through the dossier, muttering incomprehensible things. "Kelly Clark, the nutritionist. Worked in London, traveled a lot, some anti-junk food convention in Vienna. Yeah, the four-star organic-mental dingus at Kahlenberg. Security cameras show her entering her room at 10:16 pm. And never coming out. But the next day, the room was empty."

"Was that on Rue Morgue?" Alice asked.

"What?"

"Aah, forget it. Edgar Allen Poe. First crime story ever."

The Commissioner hid behind his display. Literature wasn't his preference. "Well, the patio door was open. She must have climbed down outside, or someone threw her down."

"Looks like the perp knew about video surveillance. An IT pro, a nerd who plans everything down to the last detail."

"What do we know about the procedure?"

"We believe the victims were abducted. Medichip logs show all the victims' heart rates significantly elevated some time before the crime."

"Maybe they weren't abductions. My heart rate also goes up when I see you unexpectedly."

She didn't respond. After a pause for embarrassment, regretting his idiotic remark, he asked, "What about geotracking?"

"That's even stranger. The Galileo data shows no change in location."

"Just before the murders, the chips fail. It's also interesting that we only found the unknown DNA on the victims, not on the items left behind."

"The killer wore a protective suit. Maybe the victims fought back during transport and the suit was damaged."

"Or—this is gonna sound crazy, but did you see the report on the robot army? They use it for border patrol. Maybe it's a Robby behind it."

"I like crazy hypotheses. And I'm sure EUROFORCE isn't telling us everything. Suppose the Terminator abducts the victims, but the killer is human. And he's not in the sequence database. So he's not from Europe. It's not very conclusive, but what the hell with our data situation. What else?"

"The victims didn't know each other, as far as we know."

They were pondering … the decrepit fan rattling in their minds. Air conditioning was only found in the ministry, not in the subordinate agencies.

Everything went to hell. The restrictions, the budget cuts, and one case where they were stalling. The heat wave, the damn tiger mosquitoes, Lassa fever in Europe, Finnish red wine—I shouldn't have postponed my summer holiday in Hammerfest, Storm thought.

"We have two victims who work in the organics sector," he noted. "Three if we include the biologist."

"So maybe the perp is someone who's fed up with the organic stuff." Her voice brought him back from his frustration. She was the only ray of hope here.

"Sure. There's a lot of people who hate that stuff. Pick me."

He held his wrists out for the handcuffs and regretted it at the same moment. She rolled her eyes in feigned despair.

Three months ago Storm had been appointed Chief Inspector of the Special Task Force called *Headless*, officially because one of the victims, Otto Freissler, had been tracked down by him years ago and put behind bars. The head of a gang that supplied weapons to resistance fighters in Ireland. The man had spent three years in prison. It was suspected that he was the victim of a gang war in which the opponents fought each other with bizarre rituals. The bombs matched those used by both the Irish resistance fighters and their suppliers. It was hoped that Storm's experiences with the victim's entourage would provide new insights into the serial killings, but apart from a further link to arms dealers involved in the Latin American civil wars, nothing came of it.

In fact, Storm had asked to be transferred to Northern Europe, preferably to Sweden, away from Vienna, before his past caught up with him. A past that had to do with the arms dealer in question, with cocaine, and with therapy. He was now clean, but for his former colleagues, there would always be a stigma attached to him. It came to pass that a new leader for the Special Task Force *Headless* was being sought. For Storm it was a stopover on his escape from the past.

Alice had inquired about him in advance. She had been the active head of *Headless* until the departure of Storm's predecessor, who had only been waiting for retirement. She was afraid that the newcomer would mess up her work, and she couldn't stand that. She loved her freedom and was used to making her own decisions. His dossier revealed that he had an above-average success rate in solving relationship killings. His colleagues described him professionally as persistent to obsessive, and socially closed. He had an excellent reputation as a sniffer dog who made crazy guesses. Up to now, he hadn't made any good guesses about the *Headless* case, but what could he do after such a short time, especially considering the fact that they hadn't found out anything useful in three years? The most important thing was that he accepted her as chief investigator and let her work as she wanted. He played the role of listener and keyword provider—a satisfactory cooperation, she found.

Storm shut down the computer. The panorama screen on the wall turned off.

"I've had enough for today. Want to come for a drink? Maybe the Gulf Stream will start turning sooner."

Alice was tall and slim, dark short hair combed across her forehead. Guarded amber eyes. Attractive by any standards, as well as his. An affair was not out of the question. Fragments of his daydream shot through his head. When the case was solved, he could leave again. The question was how to make her understand. She had been careful and reserved until now. Maybe it was his sarcasm that was rarely well received.

Their conversation didn't stop after one drink. It was the first time since he ran the Task Force that she revealed a little of herself. She talked about her bachelor's degree in forensics, the unexpected opportunity to work in homicide, then the first serial murder three years ago, after which the former boss had surprisingly appointed her as the main investigator; the frustration due to the lack of success in the investigation, contrasting so strikingly with everything else in her professional life, which was clear-cut, unspectacular and mundane. She had a sister who had moved north with her mother years ago to escape the heat. When she went away on holiday, she spent it like so many in the north, often with her family. She took the train to Oslo, because a plane ticket was prohibitively expensive and cost ecopoints. She had no partner; occasional affairs seemed to be enough for her, as he concluded from casual hints from her colleagues. She was considered disciplined. She talked in a conversational tone as if she was not talking about herself but about a good friend. *She's tough*, he thought. Six cases in three years. She hadn't actually made a step forward, and yet she seemed ambitious and determined.

He, for his part, told her about his law studies, the job at the ministry, the advanced training courses in criminology, some successes in murder investigations. He was sure that she already knew this from his personnel file. He didn't mention his student days, the wild parties, his only love Carol, the deep depression, and his crash after her disappearance. Actually, Alice hadn't revealed anything he didn't already know, but he still felt he'd learned something about her. She was a disciplined, persistent, attractive woman who wasn't averse to a drink after work.

The next day it turned out that the first victim, the architect, was a specialist in timber construction. Alice perked up at the news. "Now we have four murders connected to organics."

"Why four?"

"Well, wood is organic, right?"

Storm sighed. "Actually, a whole lot of things are organic. Even concrete, oil, and plastic."

"To me it's significant," Alice said.

"And the biologist?"

"Marine biology. Algae as a source of protein. Real organic, vegan, etc."

After a pause, Storm jumped to a momentous conclusion. "So we're looking for someone who hates organics so much that he kills anyone involved in producing them."

"Maybe the killer doesn't hate organic products, but the people who promote them. Maybe he's been talked into an ineffective organic cancer treatment, or his kid had developed a vegan protein deficiency."

That wasn't enough. They had to know more about the victims' environment.

"Did you interrogate the acquaintances of this Marie Rückert?" Storm asked. "I need all the transcripts."

"We scanned her private and professional life. There are two work colleagues she had relatively good contact with, and an acquaintance who was pretty messed up when he found out; probably the boyfriend. The transcripts are in the file."

"Parents?"

"The mother is dead. The father and a cousin are on the to-do list. And then, of course, there are residents, shops, bars and restaurants in the area. It'll take some time."

"Good. Would you please give me an analysis of the statements we have? I'm going to visit her father."

## Episode 2 Sweet Memories

#No kidding?

Microplastic rehabilitated! It lowers the sea temperature by almost one degree, reflecting the heat of the sun, Canadian researchers say. Waves and feelings are rising because Canada now dumps shredded plastic in the North Atlantic.

#AliceWonders

The armed conflicts on the Polish–Russian border escalate. The Ministry of
Peace reports heavy losses by Polish militias, EUROFORCE troops have
been bombed. ESA denies rumors of a project EXODUS to evacuate
important decision-makers to the Clarke moon base in case of war.


Marie Rückert's father lived in a suburb. Staff cars were only available in
acute cases, so Storm took the nearest subway. He crossed the Naschmarkt,
that run down former belly of the city, which was no longer even attractive to
tourists, leaving behind the gold-crowned Secession, the exhibition hall of the
avant-gardists of the early twentieth century. He always found the chiseled
dome roof inappropriate and kitschy, and yet he admired the urge for freedom,
the spirit of optimism that had permeated everything at that time. But all that
was over now, buried under rules and decency.

He got off at the last stop. A drone hovered over the station. The times were
restless, so surveillance was the order of the day. Storm questioned his
smartphone. It was a twenty-minute walk to Rückert's residence. That's how
much he walked each day. He was well within his typical mileage. If he kept
going like this, they'd raise his health insurance premiums.

On the way, he met a protest march of the doomsday people. The leader of
the procession held a crucifixus automaton close to his chest. Serious faces,
mutterings, crosses. The crowd moved as if through invisible mist, a sad line
of people. The crucified robot's face was distorted by simulated pain.

"Behold the signs, Armageddon is near! Repent and pray. For nothing will
remain but love." The robot spoke in a deep voice that did not fit the tiny size
of the machine at all.

Four young people came to confront them. They wore the Thunberg adepts'
sign of hope on their linen jackets. Their leader planted himself in front of the
crucifix bearer. The parade came to a halt.

"Why are you people talking about love? Love is for fucking. You want to
do something good? Save the planet!"

The crucified puppet looked down sadly at the planet saver.

"The Lord gave, the Lord hath taken away."

"What a shit!" The boy began to jockey with the cross-bearer. He and his
friends tried to snatch the praying machine. Soon the formerly peaceful dem-
onstrators pushed the young people to the ground and started kicking them.
Storm ran toward them and tried to stop the kickers. When they didn't react,
not even when he called "Police!" several times, he intervened in the fight by
trying to break up the tangle. Several men dragged him to the ground. With

a sideways roll he freed himself, pulled his service weapon while he lay on the ground, and fired into the air.

Everyone froze. This allowed the Commissioner to get up and show his police badge.

"I should arrest you all, you idiots!"

He looked around. Apart from a few bloody noses, there was no serious damage. A military drone hovered over them.

"Does anyone wish to press charges?"

Trespassing silence. Storm looked at the sorry bunch to come to a decision. Actually, everyone was in a hurry to leave. *You should be evacuated to the moon*, he thought, *along with all the power-hungry decision-makers*. The world would be a better place. Maybe there was some truth in the rumors about the EXODUS project, which were talked about in the yellow press. Sometimes he wished that conspiracy theories would become reality.

"Okay, here's what we'll do. The demo is moving on, and you"—he pointed at the four teenagers—"take care of the planet, dammit!"

And that's what happened. The machine muttered as it moved on: "Behold, I will heal them and make them well." It sounded almost defiant.

The demos increased as the circumstances became more difficult. The summers became hotter, the fields withered, heavy rain and hurricanes destroyed the harvest, the bio-wave had ruined many farmers because the consumers' demand to abstain from chemicals had been met by pests. Plastic was replaced by cotton, whose cultivation endangered the planet's water resources. The Thunberg adepts believed that alternative consumption was not the solution, but rather renouncing consumption.

They should have read Keynes. Consumption renunciation drove small businesses to ruin, because only the very big ones were able to cover their costs. An army of more and more unemployed was keeping itself alive with basic income, as the state emergency aid was cynically called, and the European Central Bank was printing gambling money to slow down the spiral of economic collapse. The limited mobility had stopped the dreaded SARS pandemics, but Lassa fever and Leishmaniasis were rampant in Europe. The vaccination skepticism that flared up again during the Great Confusion had massively reduced herd protection. Paradoxically, people were popping pills more than ever, and the strongest drugs of last resort failed against resistant super germs. Life expectancy declined worldwide, especially in the US. Mankind was suffering.

A migration of populations to the north was underway, and the deserted areas in the south were roamed by armed gangs to plunder what was left. The Schengen external border was secured by barbed wire and mines. The historic

Dublin Agreement had been reinstated and 95% of all asylum applications were rejected, with a very generous interpretation of Article 14 of the UN Charter of Human Rights. Then there was the war over Ireland, which deprived Europe of its last resources. And now the incidents on the Polish–Russian border. Fortress Europe, sad stronghold of inhumanity, misery, and despair.

Storm took a detour through a park, it was cooler there. On the yellow lawn—water was too precious for the green luxury—people sat or lay in the shade of the trees. The drone could still be heard. It hovered motionless high above the trees. When he left the park, it followed him. He tried to hide in the entrance of a house. The buzzing of the quadricopter became louder as it circled over the house. Storm stepped out and, after a quick glance up, followed his path, ignoring the drone, which quickly drifted away and disappeared westward toward the barracks. He had memorized the device number. Was he under surveillance? There was also a drone circling over the subway station...

Rückert lived in a shabby row house complex. A vegetable patch in the tiny front garden, next to the plastic bioreactor and a phosphate recycler that smelled like piss. This brought climate points. The inspector held his badge in front of the camera.

A tall, gray-haired man in his 60s asked him to come in. He was wearing a T-shirt with sweat stains under his armpits. Inside, it smelled musty of old furniture and sobriety. The living room was darkened, the air stale and hot. They sat down.

Rückert asked what he could offer. The inspector asked for water and opened his shirt collar. Took a sip of water. Just waited.

"The recycler. I have to recover it. It smells a bit strong," Rückert justified himself, as if the request for water had rebuked him.

*You said it,* Storm thought.

"You know, I can't afford fertilizer—at those prices."

His daughter's death didn't seem to affect Rückert much.

After a break, the inspector expressed his condolences and began questioning the man. Yes, he had had contact with his daughter, rarely—three or four times a year she visited him. She had a boyfriend, as far as he knew, for a year or two, a stable relationship. Controversy, jealousy? No. She never complained. Marie's financial circumstances seemed modest but adequate. She was open-minded, committed to environmental projects.

"Was she involved in esoterics? Or religious fundamentalism?"

The man sat up, barely perceptibly. "What do you mean by fundamentalism?"

"Sects, end-timers, violent Evangelicals, groups willing to break the principles of democratic order for their beliefs."

Rückert laboriously scratched his nose. Storm said to himself, *be careful. This seems like a sensitive subject.*

"Of course not. She wanted to help people. In her sessions, she tried to restore their out-of-balance mental energies. You know, climate change." He nodded meaningfully.

Storm sighed in thought and prepared himself for nonsense.

"What were these sessions?"

"In her private circle. Her boyfriend is a psychologist who specialized in mental energy flow. He worked out the scientific basis."

"Were you present at the sessions?"

"I went once. It was amazing. I felt so much better. The broken energy lines mend during the exercises. You can feel it really flowing."

"Flowing... I can imagine." Bullshit, he thought. "Did you know the people who participated? Any names?"

Rückert shook his head.

"Unfortunately, I was a stranger somehow. They were all Marie's age. I couldn't talk to them. I could barely understand them."

He was losing himself in memories of his daughter. Storm gave him time before he asked: "When did Marie develop this virtue of wanting to help people?"

"Since she was a child, actually. It started with the animals."

"In what way is that related to people?"

"Well, eating animals is not only unethical, it's bad for your health."

Marie's father hesitated before adding: "For a while, she was pretty radical."

"When was that?"

"Well, it must have been—it was 10, 12 years ago. You know, everybody was dissatisfied, the call for a new order, it had to be different, but nobody really knew where to go."

"You mean the Great Confusion? The cyber attacks on power plants, the political assassinations, the retirees pogroms?

"Yes, but for God's sake, she had nothing to do with that!"

"Then how was she radical?"

Rückert was silent.

"How radical? In thought, word, or deed?"

The man took a deep breath. "I promised not to talk about it."

"Would it harm her reputation?"

Hesitation. "Membership in certain organizations was a crime in those days."

"It was a crazy time. I was in one myself." Storm was trying to motivate the man. "But you know that after the Great Confusion there was an amnesty."

Rückert was obviously struggling with his conscience.

"Mr. Rückert, every piece of information is important. We are not investigating your daughter. We are trying to track down her murderer."

Rückert needed a long time to decide.

"It's like this," he began. "She had just quit her job in the restaurant, it was simply too much meat for her. And—okay, if it helps you: she was a member of the RAVEs."

RAVEs—radical vegans. Storm remembered the hype.

"How long was she in?"

"Maybe half a year. Soon they became too radical for her."

"Why?"

"Well, there were attacks on farmers and producers. And I think she realized that there's no point in freeing chicken or not milking cows. With age, you get insightful."

The inspector exchanged a few more petty words with Rückerrt and said goodbye. On the way back to the subway, he took stock of the situation. Hardly anything useful. The contact with the RAVEs was more than 10 years old. It was quite hopeless to get information from such a distant past, especially since the group no longer existed. But maybe the hint to the esoteric scene revealed something. They had to question the victim's friend, this psycho-energy geek, more closely.

Back at the office, he told Alice about Marie's connection to the RAVEs.

"That was cool 10 years ago," she said. "Back then everything different was cool. It was the Great Confusion. The RAVEs carried out attacks on slaughterhouses, freed chicken from farms, poisoned meat products in supermarkets, and stuff like that. It was pretty insane. But the hype is long gone. Now we have the Doomsday folks."

Storm scratched his head. "Chicken liberation. How do you do that?"

"I don't think they thought it through. The chicken were confused by their newly won freedom. There were some funny articles in the press. And the dogs had a good time. Do you think there's a motive?"

"Well, late revenge of a chicken farmer?"

She laughed out loud, then summarized the interviews with Marie's work colleagues and her boyfriend. They described the victim as friendly, helpful, politically inclined. There was no evidence of foul play or illegal activity. Her only quirk, if you want to call her that, was her trend toward dubious psychological healing doctrines, which was due to the influence of her boyfriend, who made a lot of money in the esoteric field as an energy guru. The Commissioner instructed his employees to check him thoroughly.

Later, a call came in from the forensics department. With the ceramic fragments from the latest murder case, it was possible to reconstruct the shape of the bombs. On the wall screen appeared an irregular cylinder of glazed ceramic, roughly four centimeters in diameter and twenty centimeters long, as the inserted scale showed. One end was sharp-pointed, obviously missing a piece, the other end was bent up like a funnel. The surface showed red and black paint—irregular spots that looked like snipped letters.

He stared at the reconstruction for a long time. A vase or an art object. It seemed strange to him, strangely familiar, as if he had seen something like it before. In fact, he had expected it to be in the shape of a grenade. He called the forensic expert back immediately.

"Great work, colleague. What's with the painting? It doesn't make sense."

"You got that right. This cylinder is the ultimate success with data processing. The reconstruction was only possible by putting all usable fragments of all bombs through some pretty clever software. What you see there is a mélange of the few remaining fragments of all the cylinders we found. We cannot decipher the writing, but one thing is for sure: These are neither Chinese pictograms nor Egyptian hieroglyphics. They are letters or numbers, we have identified an "Sw...," nothing more. And there are only these two colors. That's all we could reconstruct."

Storm thanked Alice and raised his thumb in a sign of victory. He invited her for a drink after work, but she declined—she had already collected enough alcopoints this month and wanted to save the remaining ones for her birthday.

Investigator's bad luck. No sex in the pipeline. Still, he would have liked to get to know his colleague, learn more about her private life, as he always did with new colleagues when he was called in on a complicated case (he called it inverse profiling). He drank the drink that she had refused, and then had more, becoming just foggy enough with philosophical digressions to find that he was hopelessly defriended, which in a way came across as very chic. In the end, out of boredom, he watched a special porn, in which he could indulge his fetish.

After that, he started dreaming weird dreams. He was invited with Alice to a banquet in a castle. They were dining on Ming dynasty porcelain, and there were delicate vases at the table. He tried in vain to cut the tough meat. All the others had eaten and were waiting for him to toast. He sabered at his steak, and half of it shot off the plate and slid across the tablecloth; desperately he tried to stop it with his elbow, without putting the cutlery down, his arm bumping into the Ming vase, which tilted as if in slow motion, rolled over the edge of the table and shattered into a thousand splinters on the floor. Alice, who was sitting opposite him, but who was not Alice, but was only wearing her smiling mask, stretched out her arm, and the splinters floated up, gathered in her presented palm to form a bent cylinder, which straightened up in her hand and swelled into an erect ceramic penis with red and black lettering. When he woke up he saw her still smiling, but it wasn't Alice sitting there, the mask fell off, it was Carol who offered him the toy.

He was immediately wide awake. Suddenly he remembered. He staggered to his feet and stumbled down the stairs into the garage. Somewhere down there must be an old box from his student days. He found it behind the folded workbench under old jeans and worn shoes. They had cost a fortune. The label had been hype, two hours in the outlet for ordinary sneakers. Crazy times. On top of a police uniform he had bought at the flea market for a students' prank, pedantically folded. Underneath, empty whisky bottles, a hookah, old-fashioned VR glasses, rolled-up posters of starlets, a football, a hockey stick, photos of jolly student gatherings. And there it was, Carol's self-potted dildo. She'd made him watch her relaxation exercises, as she called them. Maybe that's when his penchant for sex toys started. He carefully took it out.

*Sweet memories, my love* was written on it in red. She'd given it to him as a goodbye present. A farewell to sex and intimacy. A farewell also to a strangely happy, raptured, confused time.

That was 20 years ago. How could he have forgotten? But he hadn't forgotten. It had always been there, he had always remembered images of passionate embraces, romance, and sex. The images were there, but they had faded over the years. It was as if a long time ago he had read a touching story, seen a novel, a play, and only melancholy had remained.

A student affair, an ecstatic brief passion from which one lived a life. It had lasted one semester, until the hot summer (at that time they thought it was unusually hot) that changed everything. That was when Carol had met Oliver Storm's friend Peter Zigmund. It had been at one of his exuberant parties with friends, on the farm of Peter's grandparents. Zigmund and he had been both friends and competitors, not only in terms of Carol's affection. Zigmund had impressed him with his self-confidence, determination, and arrogant discipline, but he had never really liked him (for that very reason). On the other hand, their arguments and skirmishes had been cool and crazy. After that summer, Carol moved to Zigmund's farm. She and Storm had seen each other occasionally afterward—meetings with fellow students that became increasingly rare, but until the end, he had felt that there was something connecting them, if only the memory of beautiful, breathless moments.

After studying law, Storm took a job in the Ministry of Justice. Zigmund spent a few more years studying medicine, and Carol farmed, pottered, sold ceramics and organic produce. After that, they lost sight of each other. Storm learned from old acquaintances that Carol was rumored to be looking after not only the farm but also men from the village, more intensively than Zigmund would've liked. And now she reappeared from the shallows of the past, in a disturbing context, making his self-image waver. The bombs had been placed in handmade dildos that looked like his souvenir. If you could trust the forensic scientists' clever software.

The next morning brought another surprise, and one could not say it was a pleasant one. Julia Vernier, the pathologist called.

"Oliver, you wanted the results. We have everything now."

"Julia, my salvation. What would I do without you?"

"Flatterer. Well, actually, it's just like the other cases. Exitus resulted from bleeding to death. The aorta abdominalis was ruptured in the area of the diaphragm, blood pressure drops to lethal levels within seconds, whether the heart is still working or is exposed to the mechanical shock of the explosion."

"Nothing new then. And the head?"

"Idem. Dissection of the fascia and spine between the sixth and seventh cervical vertebrae. Probably postmortem."

"Probably?"

"The surgical intervention was performed immediately after the explosion. Clinically, the victim was dead. Whether there was any brain activity left, I can't say."

The inspector sighed. "Okay, thanks. Let's go to the files. Let's wait for the next case."

"There is something new, though," continued the pathologist.

"Don't keep me in suspense. This isn't a detective story here!"

"We found some vaginal tissue fragments." She put photos on Storm's desk.

"What am I looking at, doctor?"

"Vaginal epithelial tissue magnified ten times. See this white irregular triangle on the right margin?"

"It looks like a fragment from the vase—I mean, the bomb."

"That's right. It's drilled into the epithelial tissue."

"So what?"

"The epithelium is lining the vagina."

"Hmm, you mean to say—"

"That the bomb exploded inside the vagina."

They remained silent in shock.

"Coincidence, perhaps?" Storm tried. "Splinters that were thrown through the abdominal wall simply to there?"

"Possibly. But there are several samples. After we discovered the shrapnel, we went back to the images from the other cases—90% of the shrapnel in vaginal or anal tissue is found in the mucosa—i.e., the lumen. The hypothesis is quite reliable."

That was not good. Now there was a sexual component to the case. The victims had been killed by bombs that resembled the early artworks of his former lover.

Now he, too, was involved in the case.

# Episode 3 MW Medical Inc.

#No kidding?
Breaking news for astro nerds: The Earth crosser Apophis II will come damned close to Earth in 147 years. (Editor's note: I don't think the news is that breaking.) The 10 km large asteroid would release an energy of one billion Hiroshima bombs in the event of an impact.

#AliceWonders
Minister Rathenau declares that the European Peacekeeping Force will be gradually withdrawn from Central Africa. The troops are to strengthen the army on the Irish–British border and force a rapid victory over Britain.

The answer to the inspector's question about the drone came quickly:

*GZ 2019-Ex52*
*UAV EU 5728 A*
*Location Vienna, Theresienkaserne*
*Pilot: Lieutenant Jens Hasselborn*

*Monitoring of an authorized protest, surveillance of a suspect for weapons use. The suspect has been identified as Inspector Oliver Storm, Koat Wien, on duty. End of surveillance.*

*General Robert Racicot*
*UAV Fleet Central Europe*
*EUROFORCE*

Storm was reassured. He had read too much into things. Hasselborn. The name reminded him of some politician who long ago won a prize—was it the Nobel Peace Award? Strange name for Vienna, he thought, until he remembered that the military drones were controlled from the headquarters in Antwerp.

The heat wave died down, it was a tolerable 38 degrees Celsius. The Commissioner instructed his colleagues to correlate all the crime scenes and times with other crimes, to investigate the victims' past and to concentrate the search for the mysterious foreign DNA on the time 20 years ago and on countries that were not networked via Interpol. He did not hold out much hope of coming across Carol in this way, as sequencing was not introduced in Europe until much later.

He did not tell them about his relationship with Carol Beauclere and the sensitive souvenir, because the fact that he owned a sex toy that looked like the explosives that tore six people apart was anything but exonerating.

He needed to know more about Carol. What happened to her after he lost sight of her? He didn't want to involve his fellow investigators, but there was someone who could tell him first-hand. His college friend and former rival Peter Zigmund.

Storm was quick to track Zigmund down. Under the pretext of an Interpol application in a cold case, an unidentified female corpse whose description and background matched Carol Beauclere's, he visited his former friend and rival. Dr. Peter Zigmund was a research director at MW Medical, a spin-off of Doctors without Borders, which was concerned with the rehabilitation of war-disabled people.

Zigmund sat in a wheelchair, which he moved with playful ease, without using his hands to help. Storm did not dare to ask what had happened. Zigmund offered him coffee. They were sitting in his spacious bright office in the north of Vienna with a magnificent view of the city, which branched out to the withered hills to the west, on which wine had once been cultivated. In the distance one of the few green islands of the city, the garden of Schönbrunn Palace with the Gloriette on the hilltop. In the south, amidst colorful architecture and the building sins of the past, the disused waste incinerator, whose chimney with the glittering knob was listed as a historical monument, behind it in the shimmering heat the needle of St. Stefan's Cathedral. At the foot of the building, the Danube. After the meanders of the Wachau and the toils of the Viennese Gate to the plains of the East on the journey to redemption in the Black Sea, it flowed as if nothing could slow it down.

Storm broke the ice. "Hey, remember the parties at your grandparents' farm?"

Zigmund turned his wheelchair toward the window. Resting his arms on the back of the chair, he folded his hands and thoughtfully put his outstretched index fingers to his lips. Lost in memories, he watched the early morning traffic on the nearby highway. The sun flashed in the windshields of the few passing Ecars. They were heading for the Weinviertel, with villages that were already half-deserted twenty years ago and that lived on the glory of bygone times, when wine-growing was still possible there.

"My grandparents... it was a great childhood out there. When they died, the farm was left uncultivated—we had the place to ourselves. God, were we drunk."

"And arguing like the scholastics. We thought we were something special!"

"Wild times. Back then we still believed in freedom. Naive, weren't we?"

"In a way, we're responsible for what happened ten years later."

Zigmund turned his wheelchair over to his desk, swaying his head in doubt. "The Great Confusion... I don't know. The social contract would have crumbled anyway. In the 2020s, the first harbingers appeared—France, Poland, Italy, everything was falling apart. Fortunately, climate change intervened, and then the first SARS pandemic. That briefly distracted the idiots before it started again. I don't think our wild times caused the chaos. It started much earlier, with the end of the Enlightenment."

"You mean it's over?"

"As you like to think. Actually, it's not the demise, but the completion. The Enlightenment has reached its goal. Look: the web, social media, unlimited access to knowledge—the claim to know the truth and to express it freely was fulfilled with Google, Wikipedia, the social media, and whatnot. Kant would rejoice. What he did not reckon on is human stupidity. Everyone knits his own truth. The Great Confusion is the unwanted daughter of the Enlightenment."

Zigmund performed a full turn with his wheelchair, a pirouette of fatalism. "Anyway, now we're spooning out the soup—restrictions, thought police, end-times mood, distrust, war against the British, de facto war against Russia. Now they're pulling troops out of Africa. You can write that continent off."

They were silent in their worries and memories. Then Storm broke the silence. "After that last summer—you and Carol lived there for a while."

"For a while, you might say." Zigmund paused. Storm waited patiently. "We fought constantly."

"But at first you were so connected. You were a perfect match."

Zigmund made a face. "At first he said ... I worshipped her like a golden calf. But after the first love story, it was over very quickly. You know how it is when reality breaks into illusions. She was just too—erratic. What should I tell you? You knew her better than I did."

*You old fox,* Storm thought. Just rub my nose in it. It was Zigmund's pathetic satisfaction that Carol had left them both.

"She always knew exactly what she wanted," Storm said, finished his espresso and suddenly adding: "Was she faithful?"

Again Zigmund folded his hands, supported his elbows on the back of the wheelchair, and put his outstretched index fingers to his lips. His wheelchair moved rhythmically back and forth as if it were a comfortable rocking chair. "Sure. The few affairs had no meaning. At least, not to me."

"Then why is she—?"

"She wanted to leave. Here she couldn't save the world. The government's measures for the environment were ridiculous, missed the mark, and the people up there on the border were too stupid and too stubborn. I guess she was right. But there was still hope in Brazil. The rainforest, poverty, corruption, a vast battlefield for savior of the world.—Anyway, she left all of a sudden."

"And after that?"

"I never heard from her again," Zigmund said dryly.

They remained silent, each preoccupied with his memories of Carol. "She left a farewell letter," Zigmund uttered.

"You mean an e-mail?"

"No. Paper! Old-fashioned, isn't it?" He pointed at the wall, wiped his forehead in embarrassment.

The inspector got up and went to the wall where a framed piece of paper was hanging over a low filing cabinet. He looked questioningly at Zigmund and took the frame off the hook without waiting for an answer. It was a glass-covered yellowed shred of A5 format paper.

> I thought you loved me. But you're just using me, you bastard. I don't want this anymore. I had a great time with you. Now I love the rain forest. I gotta get outta here. So long, and thanks for all the fish.
> Carol

Bastard... Storm was suppressing a burgeoning gloat. "I'd like to take this with me for our forensics team on the cold case. Maybe they can do something with it. Signature, DNA, whatever. But this is not an official interview. If you don't want me to do it, I'll respect that, of course."

Zigmund looked at him, swaying his head back and forth. "I don't know. That piece of paper means a lot to me. On the other hand, if you find out what happened to her, it would be a relief. Non-destructive analysis, promise me that, okay? I want it back."

Storm promised. He thanked Zigmund for the coffee and started to say goodbye. But Zigmund interrupted him.

"If you have time, I'll show you my laboratory." He turned his wheelchair around and rolled it across the room.

Curious, the Commissioner followed. As they went down a long corridor, Zigmund started to explain his work: researching brain interfaces that would restore the interrupted nerve conduction in paraplegics. They passed through another corridor, then took the elevator to the basement. On the left-hand side, a massive door closed the way. *EUROFORCE Research … Restricted area* was written in big letters.

Zigmund took the corridor to the right. It was cold down here, Storm was shivering. After twenty meters the corridor ended in front of a glazed door with the inscription MW Medical. Zigmund opened it by facial recognition and iris scan.

They were in a huge hall. Nitrogen dewars, centrifuges, glove boxes on the walls, microscopes, fans, electronic components, cables on consoles in between. Along the axis of the room, there were endless rows of tables, on top of them vessels with nutrient solution, in which fist-sized gray objects were floating. Wires and tubes connected the objects with pumps, flashing displays, and screens. Zigmund rolled to one of the tables. In front of the screen stood a structure that looked like a metal rabbit skeleton.

"This is one of our test waldos." He touched the metal rabbit.

"A what—?"

"A robot. We call them waldos after a Heinlein SciFi novel, a term that has become quite common. MW, that stands for Magic Waldo, by the way." He gave Storm a cursory glance and continued: "The software handles the basic movements of a rabbit, but it's guided by this little guy." He pointed at the furrowed gray mass.

"Is that a brain?"

"That's right, a rabbit's brain. We stimulate neurons in the temporal lobe. The rabbit sees a certain object, like this."

Zigmund used the keyboard, and a carrot appeared on the screen. The Waldo leapt forward and sniffed around the table. "Or something." The carrot disappeared, and to the left of the screen appeared the outline of a dog. The Waldo stood up, then tried to break out to the right with frantic running movements, but he ran in place because he was chained. "Our micro electrodes allow only a limited number of objects, but that is not essential. Our future patients can see, at least normally. We want to optimize the control of limbs via the efferent motor neurons. And our *Hasi I* is doing very well." He patted the Waldo and rolled on. "The biggest problem is the decrease of potentials at a still intact synapse of the right motor neurons."

They had arrived at smaller tables where several dozen metal mice were making strange movements. "It's easier with mice. To do this, we let organic semiconductors grow into the severed spinal cord"—he pointed with a pen to the lower part of the mouse brain, which was sitting on a kind of circuit board—"and tune the output for the Waldo by deep learning. After a while, the Waldos understand what they are supposed to do." He pointed at a metal mouse skeleton on the next table, which was playfully running in a rat race.

"And what will you do if this works?"

"It's a paradox. We want to help the many severely injured people who have been tied to the wheelchair so far—whether they have a cross-section or have lost one or both legs or arms. Most of these are bomb victims—terror or war, there's so much to do."

"What's the paradox?"

"Ah! MW Medical has a huge humanitarian mission here, but guess where the money for the research comes from."

Storm shrugged.

"The military. They want brain waves to remotely control waldos for warfare. Automatic soldiers, so to speak. They use our results to create more victims in a roundabout way. It's a paradox."

Zigmund abruptly turned his wheelchair around and whirred away. Storm had trouble following him. Waiting for the elevator, the inspector pointed to the door marked *EUROFORCE RESEARCH*.

"What's behind it?"

"This is a restricted military area. EUROFORCE runs its own research lab."

"And what—?"

"There are rumors of virtual fighter planes."

"Drone pilots?"

Zigmund shook his head. "The Rift Two, hyper-realistic environment. Imagine moving in this environment—touching things, feeling thin air, guiding the stick, launching missiles, feeling the acceleration and everything, as if it were real."

"Exoskeletons?"

Zigmund laughed out loud. "It's much better. They stimulate your brain. You don't even feel it."

"How do they do that?"

Zigmund shrugged. "I don't know. Top secret. I have certain fears."

He turned the wheelchair away from the door. Storm followed him with the strange feeling that something was hidden behind the military cordon that had to do with his case. They entered the elevator and rode silently to the top floor where it was friendly and bright.

"I am in good hands here," Zigmund explained, patting his wheelchair. Storm put on a forced smile.

"I—I'm sorry. How did it happen?"

"A terrorist attack. A bomb. It wasn't personal, I guess. I just got too close." He pulled his shoulders up in fake helplessness, then rolled over to the desk and positioned the wheelchair in front of the computer.

"It doesn't matter. I'll show you how far we've come."

He reached around his waistband. A cable appeared, which he plugged into the computer's USB port. Then he typed something on the keyboard and leaned on the back of the wheelchair as if he wanted to stand up. And he did, jerking and struggling for balance. With a broad smile, he stood there and handed Storm his right hand to say goodbye before he collapsed again.

"In two or three years I'll come and visit you. On foot!"

Forensic analysis of the farewell letter yielded fingerprints of Zigmund and an unknown person. Two sets of DNAs were found. One matched Zigmund's, the other was identical to the unknown DNA found at the crime scenes. A troubling trace had become a certainty: Carol Beauclere was connected to the murders. By all appearances, she was the killer. *Jealousy*, he thought absentmindedly. Not only that, but she was alive and she might not have gone far.

The investigation now focused on the time twenty years ago, when Carol disappeared. In its latest circular, the headquarters referred to a brand new tool, pompously advertised as the Forensic Time Machine. Storm's initial skepticism gave way to growing interest when he learned the details. Forensic Time Machine, FTM for short, was the latest tool in criminology; from a myriad of information—databases, land registries, construction plans, photos, surveillance videos—the FTM constructed a three-dimensional virtual reality of the past. An ideal tool for cold cases.

FTM was developed by a start-up called RealGames on behalf of the ministry. Storm arranged a date for the training. A lively young man, a Ph.D. by age and appearance, received him in a tiny office where computers and monitors were piled up.

"Ah, commissioner Storm, your reputation precedes you!" the student greeted him, introducing himself as Thomas. "I'm sure you have a hot lead in a cold case."

He grinned mischievously and took VR glasses off a shelf.

"The Rift Two," he said, as if to explain it all.

Storm perked up. He'd heard that name before. "Wait, why two?"

"A development for the Ministry. It creates a hyper-realistic gaming environment. We're the first to have permission to use it."

The commissioner didn't respond. He was remembering who'd mentioned this software just a day ago.

"It's very simple," Thomas continued. "You enter the coordinates verbally or via Galileo-map. Photos are more difficult. The system is quite slow in recognizing patterns. On the other hand, it's super-fast to set up. We scan the cloud with our own software, faster than Google."

Storm nodded, pretending to be impressed. But his thoughts were with Zigmund. How did he know about Rift two?

"You use the data glove to control the game, like in any other."

Storm nodded again, although he wasn't a game-player. Thomas looked at him expectantly, then decided to help the obviously perplexed customer.

"Okay, let me just show you. On the screen up there, you can see what's happening."

He put on a tight-fitting data glove, held his smartphone in his right hand, put on the VR glasses, and spoke into the microphone:

"Wonderland Clubbing in Vienna last night at 11 pm."

On the screen appeared a trendy club that Storm knew from passing by. Thomas stretched out his hand; on the screen the entrance door zoomed in as if he was walking toward it. A group of young people waited in front of the entrance, chatting and gesticulating; Storm could distinguish individual voices. The camera seemed to pass through them like holograms as Thomas entered the location. He raised his hand in a defensive movement and the virtual camera stopped. He turned slowly, and on the screen, the scene rotated with him. After a few seconds, he left the site through the wall, turned around and said: "Change the time: yesterday 10 am."

The neon signs went out and the camera showed a closed restaurant in daylight.

"You know Wonderland?" Thomas asked. The inspector denied.

"Quite new here. Now watch out! Time lapse, fast rewind," he said into the microphone of his smartphone. The scene changed. Day and night passed in rapid succession. Not very impressive, Storm thought. He'd seen that before.

"Ten times faster," Thomas ordered, and the pictures whirled along. The building suddenly disappeared, a gap with cranes and construction machinery appeared, then an ugly dwelling house was built from rubble. Thomas raised his hand, and the picture froze.

"We have now arrived a good ten years ago. I suppose that means something to you?"

Storm stared at the picture. He knew the building. It was related to a murder case.

"We're going in now."

They entered the building through the front door and down the stairs to the basement. Sketchy corridors and basement doors whizzed by, all in neutral gray.

"At this point, we only have the building plans, hence the gray." Then Thomas ordered, "Slow forward," and the gray images flickered until bright light illuminated the scene. The screen showed a construction site. Pneumatic hammers had pierced the concrete floor, and bones were sticking out of the ground. A human skull lay amidst pieces of concrete.

"The ice lady," Thomas commented. "She had her husband and her lover buried in concrete in the basement. It was only discovered when the house was demolished."

Storm remembered the case.

Thomas climbed up the basement stairs and stood in the hallway, facing the exit to the basement.

"Slow rewind," he commanded his smartphone. The scene changed. People went down into the basement in reverse and then came back up. A lot of gray, when the surveillance cameras hadn't put anything in the cloud, but then a woman went into the basement in reverse; a little later (in reality probably a few hours earlier) she pushed a bundle wrapped in plastic foil up the stairs. Thomas stopped the film and zoomed in on the woman. Storm recognized her, it was the ice lady.

"A flawless proof, if we had had the Time Machine back then," Thomas commented. He took the Rift off.

"Fantastic. If it wasn't for the gray, you'd think it was real."

"They're working on it. No problem for the VR programmers. You know Underworld from Real Games? Those guys have a new thing, you feel everything, amazing. Somehow it goes with radio waves straight to the brain. It's really hip."

"Weird. I don't really want my brain to be a radio receiver."

"Maybe everything is already VR!" Thomas spread his arms and laughed out loud. It was supposed to sound funny.

The commissioner confirmed on behalf of the Taskforce that he had received two pairs of Rift-two VR glasses and two data gloves and hurried off. As he left the building, dragging the two chunky glasses and the gloves in his intelligent plastic bag—on which there was a smiley face next to the inscription "20 times used!"—a tall blonde woman approached him in the foyer. She stared at the smiley with raised eyebrows and gave him an amused smile.

Storm pulled in his head embarrassed and tried to hide the plastic bag when passing by. *Pretty lady*, he thought. How right you are, it gets on my nerves too. The world is lost anyway.

# Episode 4 Underworld

#No kidding?
The thawing permafrost releases germs that were long gone. In Siberia, smallpox and anthrax pathogens have been detected. Good morning, SARS!

#AliceWonders
Upon request, the Ministry of Peace confirms that regular troops will remain stationed in Central Africa. EUROFORCE will protect the European solar power plants.

On Storm's way home there was a game shop. When he had passed it two steps, he stopped. Something had caught his attention. He turned around to look at the hologram in the window, which depicted fighting in a bizarre cave world. *Underworld—New Touch and Feel Technology* it flashed screamingly. Above the entrance was written in blood- and slime-soaked letters "Monster Games."

He entered. Dimmed lighting, a jumble of monitors, joysticks, and VR glasses on a counter. The salesman, a youngling with beard fuzz, looked at him dismissively.

"What'll it be, man?" It sounded like, "What do you want here, Grandpa?" His clientele was at least a generation younger.

"This game—Underworld. What's it about?"

"Cool, man. I've been on the job a long time, but this—it blows your mind. Immersive!"

*You can't have been on the job that long, greenhorn*, Storm thought. He pulled out his badge. "Commissioner Strom. Want to see if we can use Underworld to train our task force," he lied, as he was ashamed by the curiosity of an old grandpa. The salesman immediately changed his tone.

"Very possible, Commissioner. You're practically inside the game. Want a demo?"

Storm nodded.

"I'm Mike. Come this way." The salesman headed into the dark recesses of the boutique. Between stacks of boxes, a customer sat in an armchair, leaning back, a pair of VR glasses in front of his eyes, his head stuck in a kind of dryer hood with cables leading to a console. His arms and legs twitched like those of a dreaming cat.

The salesman leaned forward. "Hey, Ollie, come on out. That's enough for today!" When Ollie didn't answer, Mike typed something on a keyboard. The dryer hood went up, and Ollie's spastic movements died down.

"Sometimes they can't hear you. They're in too deep."

The client took off the V-Glasses. He blinked, dazed, out of his dream world.

"Shit, man. Now Slymie has escaped. I was so close!"

"The police are here. Your session is over, man!"

Ollie jumped out of the armchair. "I'm 16 years old. Want my ID—"

"Relax, kid." Storm made a calming gesture. "This has nothing to do with you. Just want to see what's up."

*What's up, that was good,* he thought as Ollie hurried to clear the seat. Storm poured himself into it and looked expectantly at the two of them discussing quietly. "Level 1... no, no canyons... let's show him something simple... bazookas... "

Mike worked on the keyboard. "So, you are now in level 1 of the Cavern Planet. The code to the power center is hidden under one of the pyramids. The bazookas will take you inside. If you get the wrong one, you go for the next one. But beware, when the bazookas are used up, you still have the Torhammer, that will only help against the lava if you are really trained—"

"Never mind," Storm interrupted. "Just get me in there!"

The dryer hood went buzzing over his head. Mike adjusted the VR goggles. They flickered, then virtual reality built up in front of Storm's eyes—an underground corridor with smoking torches on the walls. Something was roaring; it seemed to come from the front.

"He's trimming now. This is going to be tactile," he heard Mike through the rumble. Then his arms and legs tingled like ants' walking on him.

"How do I get out?"

"The button down right," Mike said in a fading voice.

"Which button?" The inspector searched for a button on the right-hand side of the armchair, but there was no armchair. He felt as if he were standing—standing in a tunnel with smoking torches. It felt amazingly real. There was something leaning against the wall that looked like a grenade launcher. Some writing appeared above it: *Bazooka* it blinked.

He considered taking the bazooka. But how could he go? He wasn't really standing there in Underworld, he was sitting. He tried to put one leg in front of the other, and indeed he was moving along the tunnel. But it wasn't walking, it was like wading through a viscous fluid. The image jerked as he approached the grenade launcher. He reached for it and the thing jumped into his hands.

He stood there helpless. Only now did he notice a red emergency stop button on the cave wall on the right. He walked toward it, and the button moved with him. If he tried, he could feel it with an outstretched arm. The ants on his palm turned into a semblance of touch.

After a few minutes, a hint or an idea of walking was enough to float along the tunnel that soon opened up into a wide cave. In the center were three roughly layered stone pyramids. *Use Bazooka*, it flashed in his VR glasses. He aimed at the middle pile and gave steady fire. It hissed and cracked as the stone debris flew away. Soon the top of the pyramid was capped. There was smoke from the opening, and then lava poured out of it and flowed over the ground toward him. Wrong target. Escape sideways, his advisor flashed.

Sideways—spontaneously he chose left. The lava lake spread out and gnawed at his right foot. A burning pain made him flinch as the microwaves from the dryer hood stimulated the corresponding nerves. Hastily he climbed up the left pyramid. Then only the right one remained as a target. He took aim and pressed the trigger. The keystone exploded with a cracking sound and then nothing happened. An oversized red glowing hammer materialized in front of him. *Bazookas exhausted. Use Torhammer*, recommended the system.

Storm dropped the useless weapon. It crashed into the rising lava lake and melted like ice. He reached for the hammer, but instead of the handle, he felt pressure like tight bandages on both hands. Use Torhammer... all he could do was smash the pyramid he was standing on. He swung out and struck. The tip broke off and the pile of stones started to move. Storm lost his balance, stumbled, and fell backward into the lava. A lukewarm wave crawled up his legs and over his back as he drowned.

*You are dead*, the software taught him before releasing him into reality.

The two guys looked at him with a mixture of curiosity and pity. *Poor bloke* was written on their faces.

"I have something else for you." Mike tried his luck. "Apophis, ultra nerdy, man. You know, this asteroid, the Earth destroyer."

Storm had read the hashtag. A heavenly bullet, crashing into Earth in 147 years. Probability: unknown. He sighed. Planning for the future was wise; speculation nonsense. He thanked Mike and Ollie for the impressive experience, promised to get in touch, and left the shop. Impressive indeed, how these radio waves or whatever it was fooled perception. Good for a game, but unsuitable for training purposes. The sensations were too imprecise in terms of position, intensity, and quality. This technique could not replace reality.

They used FTM extensively. Over the next few days, the members of the Task Force equipped with VR glasses and data gloves were seen making strange movements when they observed events in the virtual past. It worked amazingly well. The only drawback was that the further back you went, the sparser the information became and the less accurate the representation.

Storm left most time travel to colleagues. He himself dealt with the attack which Zigmund owed his wheelchair to. Date and place were quickly determined from the logs. He programmed FTM to the coordinates and started the program. The VR glasses flickered, then he immersed himself in the simulation.

He was standing on a street in the middle of the city. He knew the area. As he turned his head, he saw familiar shops and those that had since made way for others. An arrow projected into the pavement showed him the direction. He walked down the street, ECars scurried past him. A cyclist came directly toward him on the pavement. Instinctively he swerved, but too late; the bike went right through him. As he walked on, he passed a gray, contourless row of houses, where the image data was missing.

Now he stood in front of the restaurant where the attack had taken place. *Rinderwahn* was written in pompous red letters on a projecting billboard above the restaurant which had to close after the ban on animal beef. The displayed information told him that he had reached his destination. The attack would take place in five minutes.

He walked through the closed door. The interior was gray except for a niche near the large window facing the street where the surveillance camera had been looking. Three people sat at a table for four and had lively discussions. Only murmurs and hissing could be heard, the audio filters had not been able to extract sense from the background noise. On the table was a bottle of wine, glasses, cutlery, and three plates with huge steaks. Everything was blurred and granulated. Storm approached the table, hesitated before he walked through the table and through the steaks to be able to inspect the faces of the three guests better. They were of poor quality, blurred, flickering like ghosts. He thought he recognized Zigmund in the person sitting with his back to the window, but he was not sure. The superimposed time showed one minute to go. Instinctively he stepped back a few meters. The chairs were draped with fabric that covered the chair legs at the sides. A good place to hide a bomb.

The explosion engulfed everything in blinding white until the blanked camera recovered. Wafts of smoke slowly drifted through the shattered façade window, revealing the collapsed table. The three people lay motionless on the floor. Under the destroyed armchair by the window, there was a gaping, smoking hole. The bomb had exploded directly under Zigmund.

Storm stayed until the rescue arrived, then he left the restaurant right through the destroyed façade. He traveled an hour back in time and repeated the scene. Guests arrived, others left, then Zigmund arrived with his two companions. They talked with the waiter and pointed to the table by the window. The waiter shook his head, thought for a moment, looked around the restaurant. Then something important happened: he removed a reservation card from the window table and put it on another. The guests took their seats, with one of them, a little fat man, sitting by the window first. Only after a short discussion did they change seats.

A repetition of the explosion in slow motion did not bring new insights. Everything that could be found out had already been discovered by the investigators. Even the fast rewind did not yield anything. The scene changed regularly with neutral gray, because the cameras were switched off after closing. The bomb had been placed under the armchair of the unfortunate Zigmund during one of the gray periods. Strangely enough, the street cameras in front of the building showed no attempts to enter the restaurant on the nights before the attack. There was, however, a two-hour gap in the recording of the street cameras, from which the investigators concluded that the bomb had been placed during this gap, under a table that should have remained unoccupied. It was reserved—for no one. It was supposed to remain free until the thing exploded right in front of the window without causing much damage. A statement, nothing more. The waiter had unsuspectingly changed the reservation.

The next heat wave was on the horizon, and the days went by stagnantly; *Headless* made no progress. The Commissioner felt he had missed something. Things didn't fit together. Why had Carol vanished without a trace? Why did her DNA turn up in a series of murders 20 years later? The farewell letter didn't fit either; such letters were handwritten, not printed. And why did it look like he was involved in the case? Was the bombing of Zigmund connected to the case? What was the motive? What was the connection between the victims? For the umpteenth time he went over the list. An eco-architect, an IT expert, a marine biologist, a nutritionist, a criminal he had put behind bars, and the last victim, an esoterically inclined former RAVE fanatic.

According to her father, it had been 12, 13 years ago. She was 20 after she quit her job at the restaurant. Too carnal for radical vegans...

He suddenly became alert, pulled Marie Rückert's CV from the database. And there it was: six months waitressing at the restaurant *Rinderwahn*. She quit a week after the bombing.

RAVE was the connection. He got his people together.

"I need all contacts, even indirect, of our victims with the RAVEs, going back 13 years. E-mails, blogs, postings, friends, bank records, addresses, anything."

They looked at each other, at a loss.

"Get to work!" Storm focused on Alice as the other detectives dispersed. "Alice, we need to talk. I have a hunch, and I'd like your opinion."

She said nothing, waiting for him to continue.

"It's delicate. Let's discuss it over dinner. I know a good place. You don't mind animal protein, I hope."

She raised her shoulders in vague agreement. "You guessed it. If I can afford it."

His insider tip was the back room of an expensive restaurant downtown.

"High prices," she said after looking at the menu. "You're spoiling me."

"I recommend the Sirloin steak."

"Okay. But this is insane."

He leaned over and whispered, "There's no beef jerky here. It's real."

"You're kidding. They haven't had beef for years."

"Here they do. For regulars who want discretion."

The briefing degenerated into a philosophical discussion about the limitations of the good life, the anthrax thawing from the permafrost and the fear of a pandemic, the aftermath of the Great Confusion, the question of whether things were really better in the old days and whether things had to change to get better. After a fantastic steak and two bottles of Finnish red wine, they continued the analysis in Alice's apartment. Storm was sure that the murders were connected to the attack in the restaurant. But how was it connected to Carol Beauclere? He hesitated a long time about revealing his former relationship. But his persistent colleague would find out anyway. So he took the bull by the horns and told her about the parties he used to have with friends on the farm, about his summer affair, and he didn't leave out Carol's bizarre parting gift. He told her about Zigmund and about the attack in the *Rinderwahn*.

The evening was long, and they became very close.

# Episode 5 Forensic Time Machine

#No kidding?

In 2016, trendsetters Juri Milner and Stephen Hawking put the Breakthrough Starshot project on track. Now is the time! The probe Starshot III sends super images of the planet Proxima b in the constellation Centaurus, only four light-years away. The atmosphere is like on Earth. At minus 100 degrees to plus 120 degrees not exactly the perfect holiday, but at dusk it is 20 degrees, the researchers say. And you can enjoy a swim—there is plenty of water.

#AliceWonders

Recent successes with the European fusion technology are fuelling ideas for building an interstellar spaceship. Is EXODUS possibly more than a conspiracy theory?

"We've got something," Alice greeted him the next morning when he arrived late at the office.

She hardly looked up from her screen and avoided his gaze. He grabbed an espresso, shuffled to his desk, dropped into his office chair, and started up the computer without paying attention to her. *Hangover mood*, he thought. With me the red wine, with her the intimacies she regretted in daylight. No wonder, he'd behaved like an idiot. The plan to seduce Alice with an estimated 1.5 permille alcohol in his blood was naturally doomed to failure. But of course, you couldn't see that with 1.5 permille. They had been intimate, but judging by her reaction in daylight, he had blown it. Or not. Either way, the evening had been worth it. He read the report that she had sent him, deliberately slow.

"That's something. The architect was a founding member of the RAVEs. And the biologist... Hmm, a one-time donation to the RAVEs. Tax-deductible, too. Not so convincing. What about the nutritionist who disappeared from the hotel without a trace?"

"She was lecturing on algae—the future of nutrition."

"Algae again. She had a prophetic quality. Look at all the food they serve us. Meat is banned, grain is scarce due to crop failure, and you can't afford vegetables. We should emigrate, but where to?"

"Proxima b is the solution."

"What?"

"Didn't you see the pictures? The Starshot probes. The crazy project of that Russian and the famous physicist."

Storm shook his head in confusion. "Never heard of it."

"You ignoramus. That was a hundred years ago. Mini space probes have been going to the nearest star ever since. Now they've found a habitable planet."

"For real? I'm gonna buy a ticket."

Alice Falkenberg sighed desperately. He was a hopeless case.

"What about the seaweed woman?" he asked.

"There's an interesting connection: Marie can be seen twice on the vidcams in the auditorium."

"And?"

"Four of the six victims are somehow connected to the RAVEs."

"What about the others? Just tell me, I'm not quite there yet."

She shook her head imperceptibly at such incompetence.

"OK then. The IT guy, nothing, absolutely nothing. The gunny, neither. It's you he's bothering. You put him away. That's the connection."

They pondered in silence.

"So if the RAVE has nothing to do with our series of murders, then the attack on my old friend Zigmund has nothing to do with it either. Then we have to look at the other victims. But if the murders are connected to the hit, then we have a motive: Revenge on the former RAVEs."

"So why the other two?"

"I don't know. Bottom line, we have to keep looking. And if that's true, we have a chance to prevent another victim. We need all the members of the RAVEs from the time of the *Rinderwahn* bombing."

Alice held a finger in the air. "You forgot someone."

"Huh?"

"Your beguiling one. Why is she suddenly showing up, DNA-wise? And that weird story about the dildo. Disgusting."

He leaned over.

"Listen, this detail stays between us for now!" he hissed.

"Hell, yeah!" she confirmed, unnerved. After a pause, she continued: "You realize I should document this. You are involved in this case."

They would question him and then he would have to admit that he was involved with the alleged perpetrator and had a copy of the bomb casing. And the sexual context—horrifying. That it was twenty years ago, the appeal wouldn't give a damn. They would suspend him. And if they couldn't find anyone else, they would nail him to the cross—at best as a contributory offender.

"If you do this, I'm finished," he murmured.

"All right, I'm just kidding. It's between you and me. But I bet your chick is the key."

How would he find out more about Carol? The inquiry with Interpol had run aground in the sand of the Latin American civil wars twenty years ago. There was no indication that Carol had entered Brazil at the time, which didn't say anything because there were virtually no records from that time. Everything was lost in the chaos that followed. She could easily have returned to Europe twenty years later on a false passport. But what motive could she have for doing such a thing? Even if there was one, it wasn't her style.

He decided to travel back into the past. This time the system granted an ECar. The trip to the northern Weinviertel took two hours. Along the highway, on the fertile soil, there were huge air-conditioned greenhouses where plants thrived despite the heat. After leaving the motorway, he passed a few farms that were still being run, then began fallow land where irrigation was strictly prohibited.

Derelict ranches, burnt-down barns, rusty agricultural machinery. Now and then an ancient vine stretched its stunted branches toward the sun.

The old square farm still existed, sitting stubbornly in the middle of the fallow land, which had not been cultivated for more than two decades due to lack of water. The plaster had fallen off over a large area and exposed a rough brick wall. A few clouds in the bright blue sky eased the heat. It smelled of dry earth and undergrowth. Cicada singing shrilled in Storm's ears.

He parked the car in front of the big gate. There had been a hidden key. He bent down, groped for it; there it still lay under the rain barrel overgrown with thistles, dirty and rusty, but it locked. *Your chick is the key*, his colleague had claimed.

The big courtyard was empty. Here too, thistles and dried-up grass. Bees and beetles worked busily on the few flowers. On the right-hand side, the house. There they had celebrated, drunk, smoked pot and reinvented the world. On the left-hand side, the agricultural machinery and tools had been stored. In front of it, on the opposite side of the courtyard, was the stable wing with the hay barn. The place for intimate meetings. The rough wooden door was open. Inside, it was dusky. There were gray weathered bales of straw lying around, perhaps there was another one underneath on which they had made love. He poked his foot into the bale, and the gray mass disintegrated into dust. To the right of the firewall leading to the living quarters was the large brick oven in which bread was once baked. Carol had converted it into a kiln for her ceramics. Here the dildo had been baked, whose clones had killed six people. *Sweet memories, my love.*

He looked at the oven, pensive. How often had he stood here and listened to her talk about her plans? Big plans, colossal plans, and he wasn't in them. She had always known what she wanted, without regard for others. He touched the cast-iron door of the oven, entangled in memory. Everything was like then, as if time stood still. And yet, something was disturbing. A vague feeling, nothing more. He walked down the furnace from three sides. The base was made of baked bricks. At waist height a floor space, behind it the metal door to the combustion chamber. There was the vent that led into the chimney, at the bottom a tiny cleaning door. Just above the floor, on the left side, the closable opening for the fireplace. Next to it, a collapsed pile of logs. His foot hit a log, and it trickled into brown powder.

Nothing lasts forever, he thought. It only looks that way as long as you don't touch anything.

He had taken photos then—Carol posing, laughing, holding ceramic vessels. Maybe others had done the same. It was all in the cloud.

A trip with the Time Machine was due. He took the VR glasses out of the car, put them on and started the program for the current location coordinates. In rapid return, the courtyard changed with the seasons; grass, flowers, and bushes went and came, the house walls became brighter, the plaster fresher. Everything appeared choppy and interrupted with periods of sketchy gray, when the documentation of satellite data, cadastral surveys, and private photos and videos was not complete. After ten years into the past, he stopped the return flow and re-entered the stable wing. Now he went back in time more slowly until a picture replaced the gray of a rough sketch: there was the oven, just like today. Only the stack of wood had not yet collapsed. Further back in time—much gray, if the system did not find any information. Sometimes unknown people flashed up for fractions of a second. More and more gray, the further back in time he traveled, again a shadowy flashing person. He slowed down the process, and there was his old photo that once had gone to the cloud: Carol laughing in front of the stove, showing the Victory sign with her left hand. The photo was unusually sharp. He took a closer look at the virtual oven from the past. Something seemed to have changed.

He stared at the thing, walking back and forth in time. At first he was unsure, it was an almost imperceptible change, and everything was blurred, as if the software couldn't make up its mind. It didn't help to get closer, either.

Then he saw it. The fire door was once hinged on the left, later on the right. And it was shifted imperceptibly.

He scrolled back to the old photo of himself, then moved forward slowly. Shortly afterward another photo spread—Carol and Zigmund running around in the barn, hugging, kissing, laughing. The software had converted the series to a pseudo video. He stopped, looked at the oven: just like before. Then came a lot of gray, just shadows. Later a new picture—the barn was tidy, he could see Zigmund bringing in wood with a wheelbarrow, which he stacked in front of the stove. A surveillance camera had been installed. Not much was happening. Unknown visitors came into the picture, disappeared again. He scrolled back to the last scene with Carol and Zigmund, walked close to the stove, and looked at the base thoroughly. The furnace door was hinged on the left. Slowly forward with a lot of gray, a break, until he saw Zigmund bringing in wood. He noted the time, stopped the forward motion, approached the stove again: now the furnace door was hinged on the right and offset. He took off his glasses and looked closely at the stove, put the glasses back on. There was no doubt about it: the firing door had been moved, and Zigmund had tried to hide the base behind a pile of bentwood. This was precisely at the time Carol had allegedly disappeared to Brazil.

Storm leaned against the stove. The pedestal was big enough for a corpse.

The pieces of the puzzle came together. Carol had disappeared, but not to Brazil. She'd been in there for 20 years. In the foundation of an oven in an abandoned square yard owned by Zigmund. He had liquidated his chick— out of jealousy or revenge for her infidelities. Her DNA from the serial killings was a false trail to exonerate Zigmund. As a medical man, he knew how to preserve DNA samples for decades. So Zigmund was behind the serial killings. Revenge was the overall motive. Revenge on his unfaithful lover, revenge on the RAVEs. The beautifully designed ceramic bombs were a medieval mirror punishment for those who had shattered his pelvis.

A low humming sound brought Storm back from his considerations. *Flies*, he thought and stepped out into the yard. The buzzing grew louder, but he saw nothing. Only now did he notice that he was still wearing his glasses. He took them off, and suddenly bright sunlight broke into his perception. A military drone landed in the middle of the square courtyard. He narrowed his eyes to see better. It was a large drone painted with camouflage. As the propellers died, the drone released a robot that slowly approached. The sign of Asclepius identified it as a medibot. The four arms waved like tentacles.

"Hey, what's wrong? Why is the military here?"

"Good afternoon, sir. Civil services are overloaded. I have orders to investigate you."

"There must be a mistake. I'm all right."

"Your medichip has reported a rapid drop in blood pressure. May I examine you?" the bot replied friendly.

A tentacle came dangerously close to his shoulder. It was holding a syringe.

"You're examining me with a syringe? What's in there?"

"A harmless sedative."

The fourth tentacle was digging in the instrument drawer and came up with a ceramic cylinder, which was not a cylinder, but a dildo inscribed in red and black. Two more tentacles reached for Storm. He jumped back, pulled his Glock.

"Your blood pressure is rising alarmingly. Stop moving. We will help you."

Storm stood with his back against the wall. He aimed at the tentacle with the ceramic cylinder and emptied the magazine.

And then the world disappeared into fire and pain.

# Episode 6 Ad Astra!

*Hover. A tunnel of light, longing, flickering, so far, so far.*
*It was alone. It yearned for light. It wanted to live.*

*It perceived itself, but there was nothing else. No body, no pain, no recognition.*
*Just floating and the light. Roaring, humming, vibrating. The voice of God?*
*"It's not time yet," said God or whoever. "I will make you arms and legs. A body.*
*A fine body. Be patient. You must sleep."*

And again the nightmare—the Kraken came closer. Between the tentacles was a black phallus. Oliver tried to escape, but the ground was slippery as ice. He fell, and the tentacles enveloped him, squeezing the air out of him. The Kraken tried to penetrate him, he felt the phallus pressing against him. He wanted to scream, but he had no air left.

And then the world disappeared into fire and pain.

The remnants of the dream were razor sharp, as sometimes happens when you're torn from sleep. Storm looked at the clock on his bedside table, but there was no clock. And there was no bedside table.

He wasn't at home. A strange bed, dimmed lighting, a dead TV screen on the wall, a plain table. A monitor on the left. Lines ran across the screen. *Bip—bip* it sounded to the rhythm of his heartbeat. He sat up carefully, remained seated on the edge of the bed. He felt unusually light, almost as if he was floating.

The screen woke up.

A nurse smiled at him. "welcome, pilgrim," she said.

"Hello," he croaked back. "Where am I?" His voice sounded damaged.

"You are in the hospital. You were badly injured. Welcome back to life!"

"What—?"

"Do you remember anything?"

He closed his eyes. The nightmare images exploded in his head again, but the octopus was now made of metal. Shiny metal. And the Glock.

"I fired. The explosion—"

"You were less than two meters away from the bomb. It was difficult to patch you up."

He moved his arms, looked at his hands, touched his neck, chest, and thighs. He felt himself, but as a stranger to himself. The skin on his hands was strangely smooth, like plastic. The birthmark on his right forearm was missing.

"Your skin was almost completely burned."

He moved his fingers in confusion. "Are those prosthetics?"

She denied with a smile. "The reconstruction program used your stem cell depot to repair the damaged organs. The skin will seem strange to you, it's brand new. Nerves have to work their way through it first."

He remembered an accident many years ago. He had damaged a nerve in his left elbow. For months, he suffered from strange sensations—imagined insects on his fingers, coldness when he touched a glass, and the impression that he held a stick or rope when in reality his hand was empty.

He shook his head in disbelief. He had survived the explosion. Not undamaged, but he was restored.

"Where am I?" he asked again.

"You are in the medical section of EXODUS II."

Biblical images flashing, exodus from Egypt, an old film, Israel, … then self-preservation took over.

"Was I in a coma for long?"

She hesitated to answer.

"A long time."

Since he didn't say anything, she continued, "You must be hungry. We'll send the quartermaster out with breakfast. She will answer your questions. Many things will be new to you. We want you to take it slowly. It is best to lie down again. It wouldn't be good to get up right away after so much time in the tank."

The screen went blank.

The tank? *Pilgrim* she had called him. Was this a Sci-Fi film? He was very tired and lay down on the cot. The material was soft and supple. The feeling of floating again. He looked for a blanket, but there wasn't one. He was naked, but not hot or cold. The tiny room was perfectly air-conditioned.

He turned to the side to see what bodily functions the monitor showed, on which several lines crawled, heart rate, blood pressure, respiratory function, …

The clatter of cutlery woke him. A dark-haired woman smiled down at him. Green eyes, military short haircut, the eyebrows were trimmed to a centimeter-wide stripe at the highest point of the brow arch. She was wearing a cream-colored uniform with a banded collar.

"Good morning," she said. "I am Padmé."

*She's kidding me*, he thought.

"Hello, Princess. I am Anakin."

She smiled, then her eyes got serious. She examined his naked body, nodded barely noticeably, and her face expressed all kinds of things or nothing. He sat up and covered his sex with his hands, knowing that this was ridiculous considering the situation.

She placed the tray on the tiny table, opened the cupboard, and threw some clothes at him. Then she stepped back, crossed her arms and waited. Uncertain, as she kept staring at him, he got up and slipped into the clothes. A strange feeling on his skin. As he put his trousers on, he noticed a thin band around his right ankle. The uniform jacket resembled hers, fine elastic fabric that fitted the body without wrinkles. He looked in vain for a zipper or buttons. It was enough to just touch the hem to the body and the jacket closed automatically.

"Fantastic clothes! How do you do that?" he asked.

"Nanotechnology. Very practical up here." She was vaguely pointing somewhere. She had well-kept hands, slender fingers. Her thumbs were unusually long.

"What's with the anklet?" He pointed to his right ankle.

"This is the tracking device. The system is recording where you are."

"Orwell would have loved that."

She smiled without obligation.

"It's for security. The Medibot will find you quickly if anything happens to you. The air is thin along the pipeline. Some people can't handle it. Or a GCR-KO."

"A what?"

"A spontaneous blackout caused by a GCR event. Galactic Cosmic Rays. We're exposed to a stream of ultra-fast particles, most of which are sucked out through the magnetic funnel, but there's enough left to shut down neurons once in a while. It acts like a blackout."

He didn't understand, but he didn't dare ask any more questions. He sat on the edge of the bed and waited. So many questions... where would he start?

"Bon appétit!" she said, pointing to the tray. A glass of fruit juice, a cup of coffee, light green indefinable cookies, and a bowl of white cream. He sipped the coffee, though he was neither thirsty nor hungry. The coffee smelled of nothing. He wrinkled his nose, stirred, and drank the tasteless broth. She watched him carefully.

"The sense of smell is affected by hibernation," she explained. "It'll pass."

He ate a cookie that tasted of the sea and drank the sweet fruit juice.

"Algae?" he asked, pointing to the remaining biscuits.

"Modified seaweed. It's got everything in it you need. They grow fast in low gravity."

Algae, the food of the future. Silke Eckermann, the biologist. He took the biscuits slowly and drank the brown broth in between. He watched her watching him.

"How long have I been away?" he finally asked straight out.

"One hundred and thirty-seven years, three months and six days."

He coughed up the chewed-up piece of meat and gasped for breath.

"This is a shock for many," she said compassionately. "Come, let's take a tour, then I'll introduce you to the commander. I'll explain the most important things to you on the way."

She took him by the hand and pulled him up. He was leaning on her shoulder, otherwise he would have tripped.

"Be careful walking, gravity is low at level 1."

A huge hall spread out in front of the sickbay door. At first he thought they were in the open, but high up there was a ceiling. When he squinted, he saw something like insects crawling across a strangely curved canopy.

The floor was curved, too. It seemed as if they were inside a huge hemisphere, halfway between the pole and the equator. There, a good 50 meters away, the hall was closed off by a front of houses with balconies and terraces. Above it, supported by massive columns, hung a honeycomb pattern of steel girders, which took up the entire field of vision up to the ceiling. From the middle of the steel construction, a mighty shiny silver tube protruded halfway up. It crossed the area and disappeared into the wall of the medical department behind them. The light also came from there—the beam shone in a warm yellow tone.

"Where are we?" he finally stammered.

"We are on the interstellar ship EXODUS II on its way to Atlantis."

*She's screwing with me*, he thought.

"Atlantis," he said, feigning sobriety. "Somewhere in the Mediterranean. But doesn't really exist. So you're making a science fiction film?"

She didn't get his sarcasm. "Not at all. This is the planet you used to call Proxima b."

"You mean, the planet they recently discovered?"

"Recently? A good joke, indeed."

They crossed the square in silence. Storm walked as if on cotton wool. Padmé had to prop him up, because he staggered at every feathery step. He was gripped by an all-out perplexity. Add to that the strange feeling of always standing upright, even though the ground in front of them seemed to rise, as if they were moving in a huge hamster wheel.

Soon they stood in front of the houses. Here he felt safer on his feet; the ground seemed more stable. A mighty screen, 200 inches or more, was mounted at medium height. Written on it in big letters was:

| | |
|---|---|
| Distance to destination | 3.473 ly |
| Time to arrival | 9 y 213 d |
| Distance from Earth | 0.770 ly |
| Speed | 0.398 c |
| Acceleration | 0.105 g |
| Slip correction | 39.67 d |

Below was more text, but he couldn't read it from a distance.

"Great service. Just like on a plane. What is *ly*?"

"Light-years."

He laughed as if she had told a Dadaistic joke.

"...and other nice numbers," she continued. "Our physicist can explain everything."

"Nine more years?" he asked incredulously.

"Nine years and seven months," she confirmed. "If all goes well."

A spacious promenade lined with shrubs and flowerbeds ran parallel to the buildings that bordered the open space they had just crossed. Glass and steel everywhere, green balconies and terraces. It looked like experimental architecture that boldly curved upwards on both sides. Above the three-storey buildings, the honeycomb construction of steel girders reached into the sky. High above their heads, where the luminous truss stood out, it seemed somehow folded, hanging over them like a threatening rocky outcrop.

He pointed upwards.

"Is that the ceiling light?"

"That's the pipeline."

"Oil."

"No, the hydrogen pipeline. Fuel for the fusion drive. Listen, this is too complicated now. But if you're looking up—"

She grabbed him by the elbow, turned him around and pointed at the ceiling.

"How high do you think the ceiling is?"

"It's hard to say. Twenty meters?"

"It's ninety. And what do you see up there?"

The insects were still crawling around there.

"It looks like ants. Big ants."

"Look closer."

Now that he knew the distance, the insect hypothesis was excluded. He squinted again and now he realized they were humans, people in uniforms like they both wore. And they walked upside down.

Padmé circumscribed a circle with both arms, then pointed to the pipeline.

"This is the axis, pointing in our direction of flight. The EXODUS is constantly rotating around this axis. It's not gravity that keeps you down, it's centrifugal force. We're in the *belt* here. If you keep walking this way"—she pointed to the right along the road—"you will come back to the starting point after a full circuit."

His gaze followed the house front. The avenue curved upwards with the buildings and merged with the ceiling. The promenade was self-contained like a ring, standing vertically.

"Here on level 1, the centrifugal force is between 0.2 and 0.3 g."

"I know a few psychology laws, there is also something like centrifugal force."

"One g would be Earth gravity. Where we're standing, there's a third of the Earth's usual gravity. You only weigh—let's say 25 kilos."

He raised his hand to interrupt her and smiled skeptically, as if she were joking.

"What's really going on here?"

She sighed in half-feigned desperation, grabbed him firmly by the forearm, and lifted him up with one hand.

"Will that satisfy you, lightweight?"

She let go of him and he floated gently to the ground. He gazed up at her, stunned. Something was wrong here. This happened only in movies.

"The fusion drive of the EXODUS creates a thrust equal to one-tenth of the Earth's gravity. This is superimposed over the centrifugal force of rotation."

Since he didn't understand anything, he kept her talking.

"Level 1 is close to the axis of rotation and the centrifugal force is weak. Here, the thrust is the main force. The further we move away from the axis, the stronger the centrifugal force. It always points outwards—away from the pipeline, at right angles to the thrust force. So the sum of both forces changes direction as we move away from the axis of rotation. The habitat is curved so that the force is always perpendicular to the ground."

Storm stared at her, uncomprehending.

"Just like on Earth," she said. "The gravity is always perpendicular to the curved surface. Here it's similar, only the curvature is concave, not convex."

"You mean the whole thing is hollow? A hemisphere?"

"Not quite. It's a paraboloid."

He kept staring at her. Her expression left no doubt that she thought he was mentally handicapped.

"Do you drink coffee?"

"Too much."

"When you stir the cup, what do you see?

"The milk makes nice little circles."

"You're very observant, Inspector. But the surface, what does it look like? When you stir very quickly?"

That's when he understood. "There's a depression in the middle. Due to centrifugal force."

"Exactly. A paraboloid."

She turned around, looked over the empty space they had crossed and spread her arms. "We are in a huge rotating coffee cup," she explained radiantly.

He scanned the fronts of the houses. They formed the edge of the coffee cup, the igloo the center.

"And where is the milk?" he asked dryly. She tilted her head and presaged a smile.

On the right-hand side, there was a semicircle with rows of seats, reminiscent of an antique amphitheater.

"The arena," Padmé explained. "Our gathering place."

In the center of the stage was a column on a pedestal. Storm pointed at it.

"The flagpole?"

"That's the pillory."

"The what?"

"The pranger. Didn't you have that in your century?"

*You should read up on history*, he thought. "For your information, the Middle Ages were several hundred years before my century. How romantic to have it back."

She didn't notice his sarcasm. "Not everyone agrees with this method of punishment. But, it's the captain's decision. This is a warship, so we're not squeamish here. Back in the seafaring days, people used to get keelhauled when they did something wrong."

"I guess that procedure wouldn't be so easy here," he muttered.

On the left was a fenced-in area, green-painted ground, netting, a high referee's chair. The ground was curved concave on all sides like a monstrous shaving mirror. Two players were on the court, and with grotesque giant strides, they floated more than they ran from sideline to sideline. You could hear the dry *plock—plock* when the rackets hit the ball, which followed an impossibly curved path as if it had enormous sidespin.

"Interstellar tennis?" he asked. It was meant to sound sarcastic, but she nodded seriously.

"3D tennis. The Belt is our recreational area."

She pointed at several structures, one after the other, each extending to the ceiling on both sides in a semicircle. Open spaces with floor markings, buildings, trees in between.

"There is a handball court, two small ponds, a forest, gardens, a gymnasium, a dance floor, a theater, an art wall. The Belt is bordered by two promenades. We are on the big one. On the other side, touching the igloo, is the small one. It's closer to the axis, where the centrifugal force is less."

The small promenade was also a vertical ring that ran around the perimeter of the belt. A huge igloo-shaped structure, with the pipeline disappearing at its center, closed off the ring. There was the opening from which they had stepped out into the square after his rebirth. It had stood vertically, as doors should, he was sure of that, but now the floor and the door were tilted forward by 45 degrees or more. Storm had to force himself to think of the coffee analogy to understand. Still, it was unnatural, like a cabinet of mirrors.

"What's there?"

"Medicine, bionics, and research."

"You mean to say I was frozen there for over a century?"

"Not there. After the bomb attack, there wasn't much left of you. You were hibernated at EUROFORCE because"—she interrupted herself—"Man, are you curious. I can't explain the details to you now. It will all come later. Anyway, you're with us now, pilgrim. And you should prepare for your first shift!"

She looked at him, probing, like he was a pet. As if she were wondering whether to buy it.

"Okay, maybe I'm rushing you too much. My fault. Listen, we're changing up the plan. I'll show you to your quarters in a minute. The rookies are on level 2 with 0.5 g to get them used to their weight again. You can pull all the details on the Rift and interview the chatbot, but don't overdo it. The first steps with a new body are not without its problems. You have to connect again first. Tomorrow is another day."

Padmé's other day lasted three days. The first night had been restless—if there was actually a night here. The nightmare came again. He found himself in a no man's land between being awake and dreaming. The octopus, the explosion, the starship, the departure into an uncertain future, hibernation, Alice, Padmé, Atlantis... all blended into a confused pastiche.

In the tank he had only had a tenth of his body weight, so the descent into the living quarters had been tiring and exhausting, but after two days spent in his new apartment he found the 0.5 g in level 2 comfortable; it was as if he couldn't feel some parts of his body. Half asleep, in floating dreams he was looking down on himself as if his soul did not accept this strange reconstructed body.

The apartment was no more than a bunk. On command the couch became a chair with a little table. There was a mini-sink and a niche in the wall; that was the food serving area, Padmé explained to him during one of her short visits. There were floor showers and toilets.

The Rift, that was the VR glasses. They were hardly different from the ones he knew from the Forensic Time Machine. He liked the Rift; it was real, it was the link between two worlds. When he touched it, he was almost certain he wasn't dreaming. When he put it on, Padmé—or rather her avatar—appeared as lifelike in 3D. The heavy headband of the VR glasses vibrated, and for seconds he had a strange feeling of floating.

And Padmé's avatar explained the new world to him.

The EXODUS II was the most modern interstellar ship of the EUROFORCE battle fleet. The sister ship EXODUS I had been launched 18 months earlier. Three years after the launch, they had a problem with the fusion engine that could not be fixed. The reactor had exploded and destroyed the ship.

The EXODUS II was to take its crew—12 teams and 150 passengers, the pilgrims—to the new world. The teams worked in shifts. One 30-strong bunch was active for a month, the others were in the box, as the hibernation was casually called. There was a shift change every month, so that after a year all teams were through. There were two management teams (captain and officers), which took turns in the annual cycle. The voyage would take fourteen years; but thanks to the hibernation in the "box," each team was awake for only fourteen months. Failures, changes, or additions to the crew were covered by freshly reconstructed crewmates from the tanks.

Propulsion was provided by a Bussard ramjet. Superconducting toroidal coils on the outer hull of EXODUS generated giant magnetic fields that sucked in ionized interstellar hydrogen from a radius of hundreds of kilometers and compressed it along the axis as in a conventional ramjet, feeding it into the fusion drive. The small proportion of neutral hydrogen in the interstellar gas was ionized in the electron stripper, a proven concept from the early days of particle accelerators. The faster the ship was, the more hydrogen was harvested, and the stronger the thrust, which allowed acceleration of a maximum of 1.2 meters per square second, corresponding to an inertial force of about 12% of Earth's gravity in the axial direction. The fusion reactor was located three hundred meters behind the habitation module and was connected to it by the pipeline through which the hydrogen flowed. The temperature inside the combustion chamber was several hundred million degrees. The waste heat was radiated through huge heat sinks that were mounted along the pipeline like solar panels, called lotus leaves in the ship's jargon. A meter-thick shield of heavy elements protected the habitat from the by-products of fusion—neutrons, relativistic electrons, X-rays, and gamma radiation. Communication with Earth was maintained by means of a network of phased array antennas at the bow—the spider.

The habitat was divided into four levels with increasing centrifugal force. Level 1 at the stern housed the igloo with the medical department from which he had emerged a few days earlier, the tanks, as well as the hibernation for crew and passengers, and the genetic databases with human and animal germ material. This was followed by the Belt, the hollow paraboloid he already knew—a leisure park for relaxation and sports between the igloo and the living quarters with low centrifugal force. Since the Mars expeditions, it was known that physical training at 0.3 g was sufficient to avoid the harmful effects of weightlessness. The leisure area was limited at the bow by a complex of buildings, at the front of which the flat screen displayed the voyage data. There, in level 2, the ground floor housed the living module with the crew apartments at 0.5 g. On the upper floors, where the centrifugal force was weaker, there were offices and recreation rooms. Above this, a honeycomb-shaped steel construction stabilized the area up to the axis of rotation.

Descending from the living module to the area further away from the axis one arrived at level 3 with 0.7 gravities. It housed the bridge, telemetry, communications, and the "garden"—that was the food supply, the granary of EXODUS, in which algae, fruits, grains, and vegetables thrived under bright sunlight harvested from the fusion engine. Animal protein was cultivated in a bioreactor. Level 4 at 0.9 g on the outside included central control, recycling, energy supply, building services, storage, training rooms, and the chemical factory. The water tanks that covered the bow of the habitat served as a protective shield against cosmic radiation. On the outer wall of EXODUS, protected from cosmic radiation by the funnel and the habitat, were the landing modules for the arrival on Atlantis and the three longboats Mayflower, Discovery, and Endeavor.

With the VR glasses Storm made a virtual journey around the spaceship. At the stern sat like a sprawling bulb the powerful engine—fusion torus, compressors, high-pressure pipes, nozzles—from which the blue-flaming reaction gas was emitted like delicate roots. Upwards toward the bow, the stem of the pipeline bored its way through the meter-thick red-hot radiation shield as if through hard soil, growing many hundreds of meters high, covered by heat sinks that sprouted all around like lotus leaves, until the stem opened into the shiny silver cup of the habitat, surrounded by the superconducting ring coils of the magnetic funnel. The open flower was surrounded by the wide veil of the communication antenna. The semitransparent main electron stripper floated above it like a filigree protective shield. When zooming in, Storm spotted tiny service bugs crawling around. An interstellar flower, thrown into the stream of time in the hope of finding fertile ground in a new home. In front of the bow, Alpha Centauri shone like a beacon that showed them the way. As if to protect them from the dazzling glow, a forest of thousands of transparent discs hung a hundred kilometers ahead in space, only recognizable by their frames—the vanguard of the electron strippers that covered the invisible edge of the magnetic funnel.

Storm slept all day. During the night his nightmare repeated, and this time after the explosion he saw himself as if his soul rose from the torn body. He looked down on himself while he floated away. In the morning he felt ravenous, consuming vast quantities of strange food. Padmé urged him to exercise his muscles, which he did well. The exercise came easily to him in the low centrifugal force in his quarters, and although he trained hard and every movement was strenuous, he had no muscle pain. His body perception slowly returned and the initial dizziness subsided. The sensation of touch was disturbing. He pinched his thigh and felt nothing. Below his knees there was a tingling sensation when he touched firmly, he felt his ankles and toes except for the small toe joints, which were absolutely numb. On the first day after waking up he had had trouble holding cutlery or glasses; this soon improved. Only the index and middle fingers of his right hand did not do what they should. He had the feeling that they were mixed up when he touched them, just like in the children's game where you had to guess what they felt with crossed fingers without looking.

The exercises gave him the opportunity to get to know his new body. He explored himself in the training video. They had reconstructed him slimmer than he had been. The skin was taut and youthful, the face also looked fresher and younger. Watching the control video for the muscle exercises he made faces, snarled, pulled his ears, bent his nose in all directions, frowned, pulled his brows together, stuck out his tongue, examined his hairline, which unfortunately had not grown any lower. It was the wrinkles, he finally realized—they hadn't disappeared, but they seemed less carved in stone than he remembered them. Well, he had been lying in a nutrient solution for over a century.

Sometimes he heard footsteps at night. They came from the floor or the walls, accompanied by murmurs that reminded him of voices muffled by water. Often he thought he recognized Alice's voice. Then he left his quarters to search for her. Obviously, he never encountered anyone. He blamed it on his exhaustion and the loss of his former life. *Ridiculous*, he reproached himself.

He had to learn how to live again. And he began to learn how to walk again. Secretly, for the time being. When he saw someone, he turned away quickly, as if he were a blind passenger on a luxury cruiser. He felt strange, as if he was slowly recovering after a high fever, when the line between being awake and having a feverish dream still blurs. At night he ventured as far as the Belt, where he trained carefully, only to sit down in the arena, exhausted after a short time. Breathless, still unable to believe where a bizarre providence had thrown him, he marveled in the dull glow of the nocturnal pipeline at this enormous construction, which, untouched by human hope and fear, was constantly turning about its axis.

The arena offered a clear view of the enormous flat screen.

| | |
|---|---|
| Distance to destination | 3.471 ly |
| Time to arrival | 9 y 211 d |
| Distance from Earth | 0.772 ly |
| Speed | 0.398 c |
| Acceleration | 0.105 g |
| Slip correction | 39.84 d |

## Episode 7 Exploration

Two days ago he had seen the screen for the first time. They were approaching their destination at a snail's pace. The lines below, which he had not been able to decipher at that time, contained the latest news from Earth:

#news flash 3.7.2214
After CIA archives were opened, it became public that the USA had withdrawn from the International Climate Agreement in 2017 with the intention of accelerating the thawing of the permafrost in Russia. This should lead to destabilization of the infrastructure and economic performance.
After Pakistan's nuclear attack on Jodhpur, India destroys Peshawar in a nuclear counterstrike.
© RG channel via moon base Clarke II, sent 26.9.2213

It took him a few minutes to process message two of the news. Earth was in a bad state of affairs.

He spent most of the day exploring the Rift, trying to learn about life aboard. In the archives, he found documents about planning and building the spaceships, animations about the long journey and books about the world political situation that had led to the project.

The century he had slept through in the reconstruction tank had been marked by war and violence. Even before the turn of the century, the average temperature of the Earth's atmosphere was already 19 degrees Celsius, three degrees higher than the Paris climate target of 2015, making it clear that global warming had already derailed and could no longer be stopped by human intervention. The worst forecasts were exceeded. Pestilence, famine, and extreme weather conditions struck the Earth. The climate system was rapidly evolving toward a scenario of a hot Earth. When he finished reading, he recapitulated the sad story:

In Europe, in the first half of the twenty-second century, an exodus of populations from Spain and southern Italy takes place. Summers are extremely hot and dry, mobility is rigorously restricted for economic and climatic reasons, unemployment rises to record levels and migratory movements across the Mediterranean become unmanageable as EUROFORCE has neither the means nor the mandate to intervene. Core Europe, the entity that emerged from the former Western European countries, is essentially abandoning Spain, retreating to the areas north of the Pyrenees, and in Italy north of the Po Valley. China is buying land on a large scale south of these lines and establishing special economic zones with autonomous status. This creates employment, by and large for the Islamic migrants and at the same time a buffer zone against Islamist terror. The southern external border of Europe is secured by Chinese troops. The eastern border is insufficiently protected.

After the chaos of the civil wars, Latin America will never find its way back to stable states. The political turmoil is preventing climate-friendly counter-measures following the massive land clearing and the decline of the Amazon rainforest. Russia is sinking into the swamps of thawing tundra and corruption, there has been a state of war between Pakistan and India for decades, China is pursuing an efficient trade policy and keeping pace technologically with Europe and the USA. Australia-New Zealand, as the country is now called, is in many respects favored by its geographical isolation and is less affected by the consequences of climate change. Europe, China, and Australia-NZL are the only nations that maintain permanent manned moon bases.

The biosphere is changing, tropical species are moving into temperate latitudes. South America is becoming drier, the rainforest replaced by steppe and grasslands, run-away effects are setting in. Nonlinear feedbacks such as the melting of the polar caps, the emission of $CO_2$ from the warmed oceans and the release of large quantities of methane in the thawing tundra lead to a tropical warm climate in the middle of the twenty-second century, as it prevailed at the end of the Paleocene 55 million years ago, with average temperatures 10 to 15 degrees higher than a century earlier. Sea levels have risen by 40 meters in the last hundred and fifty years. The earth is almost ice-free, the continents have new coastlines, and the habitable zones are above the fortieth parallel at the end of the century.

Europe withdraws to areas north of the Loire and Danube. South of this line, in the now extended autonomous areas under the Chinese protectorate, migrants live in slave-like conditions. Rebellions are crushed brutally to the secret relief of the rest of Europe. The Middle East is depopulated by religious and civil wars and has fallen back to preindustrial levels. The extreme drought in the Midwest of the USA, which is covered with huge solar farms, leads to migration to the north. Canada closes its borders. Southern California has no more water, Los Angeles and Las Vegas are ghost towns. A 100-km wide strip of no man's land up to the border fence with Mexico is roamed by dubious marshals who have carte blanche to secure the border. The USA wage an economic war against Europe and Canada. US interest in methane hydrate deposits off the Canadian East Coast leads to armed incidents.

Turkey has annexed Syria, and Iran its enemy Iraq. Israel has occupied Lebanon and the Gaza Strip and, under the Treaty of Amman, buys the West Bank from Jordan, which acts as a buffer zone to Iran and Saudi Arabia.

In a mild climate, the Siberian tundra becomes a granary after permafrost thaws. Russia prospers through grain exports to the USA and Europe.

Around 2160 the global migration begins. This leads to wars and ethnic massacres in the last third of the twenty-second century, especially in the USA against Mexicans, in the EU against Arabs and blacks, in South Africa and Namibia against whites, in India against Muslims, Christians, and Hindus, in China against Tibetans, Uyghurs, Yue, Hakka, and Wu. Diseases, famines, epidemics, and environmental disasters are on the rise.

In 2180 Israel destroys the Iranian nuclear facilities at Natans, Qom, and Arak, releasing large quantities of radioactive material. The nuclear counterstrike kills 70,000 in Tel Aviv. Israel subsequently fires Jericho missiles with nuclear warheads, two of which reach Tehran. Russia meets its obligation to provide assistance and warns against further Israeli actions with a nuclear attack in the Negev desert.

The threat of global nuclear war is mounting. Socioeconomic studies show that a relapse of mankind into a prescientific age is imminent, similar to the early Middle Ages after the fall of the Roman Empire. The EU decides to build starships for a new dawn of humanity. A secret plan of EUROFORCE has been in existence for a long time. The propulsion is based on European fusion technology.

China and Australia-New Zealand draw similar conclusions. Almost simultaneously, the three nations begin building starships in the libration points of the Earth–Moon system. Europe occupies L1 between the Earth and the Moon, shortly afterward China takes L4, and Australia-New Zealand chooses L5. Although Europe has a transport advantage by positioning in L1, China is faster. First, in 2197, the generation ship JIAN TOU takes off toward Proxima b. Planned voyage duration: 31 years. Eleven years later, the European

EXODUS I takes off with a second-generation engine. The EXODUS II follows after one and a half years with state-of-the-art technology. Both ships are to overtake the JIAN TOU and reach Atlantis ahead of her.

At the beginning of the twenty-third century, the world population has fallen to 4.2 billion.

On the third day, Storm was steady enough to be presented to the crew.

Padmé picked him up and briefed him on important details of life aboard. The commander had almost unlimited power. His orders were sacrosanct, and even civilians were required to obey them. The officers formed a small, well-coordinated hierarchical network. Formally, martial law prevailed. The basic rules were rigid, and there had been two death sentences—one for murder, the other for attempted sabotage of the fusion reactor. Minor offences were handled rather loosely. The worst that could happen would be the pillory. Storm also learned why only the non-reconstructed had trimmed eyebrows—that was trendy in the twenty-third century. And the peculiarly long thumbs?

She laughed. "That's probably when evolution struck. If the smartphone is your major organ, long thumbs offer a survival benefit."

She explained the basic social rules to him. Not much had changed; shaking hands was politically incorrect. The crew was on a first-name basis. Officers were called by their family names, as did officers to each of them.

Just before they came to the meeting, he held Padmé back.

"Listen, this may not be the right time, but I don't think there is one. Can you explain why I, of all people, am here? I mean, in case they give me a grilling, what should I say?"

"Don't worry, it's a formality," she replied. "And you'll meet people who had a similar fate. It's no problem."

"Still. Please!"

She sighed demonstratively.

"Okay, pilgrim, in fast motion," she explained. "At EUROFORCE, back when you got bombed, they were running a secret project to cryo-preserve war-disabled and bombing victims. The idea was to put severe cases into cryogenic sleep until medicine could help them."

"The bomb," he muttered. "Who attacked me?"

"It was a drone. You fired on it, which set off the explosion prematurely and saved your life."

Boiling hot his repressed memory popped up: Sweet memories... Carol's fetish, the ceramic cylinder, painted in red and black. Shrapnel bombs. The serial murders, the suspicion of Zigmund, his former friend and rival. Carol—she had not disappeared in South America but was bricked in an oven in an abandoned farm a light-year away. When Storm had discovered this, the drone had landed to eliminate a prosecutor—the octopus with a bomb in its tentacles.

Zigmund was behind it all. He had laid a false trail to hide the truth.

"Did they catch whoever did this?"

"We don't know."

"But there must be details. Investigation results, newspaper reports. Can't you google that?"

She smiled pitifully.

"We don't have a world wide web on board."

"But you have an archive. Documentaries from the past, all the knowledge of mankind..."

"We have an extensive collection of science and art. But we don't have newspaper clippings from your era, more than a century before. And certainly no police investigation files from the Antique."

*Antique*, she had said.

"Can we request that?"

"Sure. I'll include this in my next report to EUROFORCE. You'll have your files in two years."

"But..."

"Mail is slow here. We're almost a light-year away from Earth. The request arrives there in just under a year, and the answer will take a little longer since we'll have traveled on in the meantime."

Storm fell silent, locked in memories. It was all an eternity ago. A ridiculous assassination attempt, one of many, with no long-term consequences whatsoever. His case was too insignificant to be worth much notice.

"Look, it doesn't make sense. In the big picture, I'm a lightweight, a nobody. So why am I here?"

"It's simple. The committee wanted you on board."

They entered the large square on level 1, which he already knew. The arena opened up before them. The curved floor, the strangely tilted igloo of the medical department in the distance, the pipeline on the axis of rotation, above it the now deserted promenade on which three days ago people had walked upside down. Storm was gripped by a sudden longing for the unknown, for departure and adventure. His pragmatism took over. The bomb attack? That was old hat. Up and away to a new world!

The ranks of the amphitheater were occupied in a semicircle. Everyone present wore the same uniform. Curious eyes were everywhere.

Padmé showed him the way to the first row, heading for two free seats. Storm followed, jumping like a kangaroo in 0.3 g. A man in the next seat stood up.

"I am Horst Thelen, first mate. Welcome aboard the EXODUS!"

"Oliver Storm. Pleased to meet you—at least I think so."

Thelen laughed loudly and patted him on the shoulder. He too had slender hands and extra-long thumbs.

To the left of Padmé sat a thirty-year-old man with a Henri-quatre beard. He was typing on a peculiar tablet, muttering incomprehensible commands, whereupon the tablet grew in size and showed a large blueprint of the ship. Occasionally he looked up, critically examining his surroundings. Storm did not fail to notice that the bearded man was watching him stealthily. So he leaned forward to the left to address the man.

"Hey! I'm the new guy."

"Charmed. I'm Jason, the used one."

Storm looked up to Padmé for help.

"Jason Nygard," she said. "Our physicist. We don't understand him, but without him, we wouldn't understand anything."

The physicist touched the sides of the tablet, which shrank like magic, curled up, and metamorphosed into a cigar-sized cylinder. "Padmé, my dear, too much knowledge is dangerous. The Mafia knew that centuries ago."

The guy pleased Storm. "Cool thing," he said, pointing to the strange gadget.

"The mablet?" Nygard fiddled around with it like with a ballpoint. "Right, it did not exist in our century."

Padmé pointed at the houses that bordered the square. On a terrace on the upper floor, about ten meters to the left of the stage, a large sturdy man appeared, looked around as if he was making sure everyone saw him and jumped down diagonally with a playful swing. He sailed straight for Storm. "Sailing" corresponded most closely to his mode of locomotion, as he flew as if in slow motion, but then deviated in a slight curve to the right and landed in the center of the stage. People applauded.

"That's our captain," Padmé commented. The physicist shook his head and murmured dismissively.

The Captain raised his arms, and the applause died down.

"No offence, dear friends. You know how it is, and you know me and my love of sports. That was a demo for our newborn pilgrim.

In his former life, Oliver Storm was a famous Commissioner. He has risen from the dead. Yes, you could say that, because without hibernation he would have been dead for over a century. Most of you come from the tanks—so many victims from assassinations, accidents, disasters, war. They found Oliver just in time then, it didn't look good for him. The bomb did a good job."

He pulled a note from the breast pocket of his uniform, unfolded it elaborately and read it:

"Left eye missing, right arm off at the elbow, one lung torn apart, liver lacerated and—well, I don't want to read the list, it's disgusting. In short: everything below his diaphragm was bloody mud. Anyway, he was a borderline case. The doctors had little hope. The reconstruction had been going on for a year when we started. I thought long and hard about whether I should list him. But he was a good fit—his psychological parameters and know-how convinced me. And I was sure he was a nice guy. And that's why you're here with us now, Oliver Storm. Welcome to the EXODUS!"

He came up to Storm, who stood up and let the captain hug him. The crew applauded again.

"We have a job for you, Commissioner," muttered the commander during the hug. "You'd better get started."

The grumpy tone didn't match his smile. Storm did not like the man.

With that, the official part of the welcome was over. Storm was surrounded, everyone talked to him, and he heard many names he didn't remember. The captain's name was Torsten Ahlgrim, and there was a doctor called Lucas de Vries. The people resembled those he knew from his first life over a hundred years ago. Everything was as familiar as at a debut party at a forensics seminar, except the curling tablets which he spotted every now and then. De Vries eagerly explained this twenty-third-century technology. The mablets—semi-intelligent "magic" tablets—were made of a paper-thin graphene-organics compound.

It took Storm a minute to grasp why nothing else seemed to have changed in a century: They all were resurrected from the same era, except for the officers with their strangely trimmed eyebrows, which gave them an attentive expression. *That's something that should be introduced at conventions*, thought the inspector before he realized that there would be no more conventions.

"What did the Captain mean by 'most of you come from the tanks'?" he asked Padmé as they left the amphitheater. "I thought the European elite should be saved from Doomsday."

She stopped.

"On the passenger list were important persons—politicians, artists, scientists, people with special skills. They had all been carefully selected. Then came the nuclear conflict between Israel and Iran. Russia and the USA interfered. Our strategists forecasted a worldwide nuclear war. The order to launch came out of nowhere. It happened too fast, important people have important tasks, especially in times of crisis. They preferred to stay at home. A probable demise on earth is apparently better than an uncertain departure to the stars."

"Not a single Nobel Prize winner? Scientists, politicians, artists?"

"Sure, Greta Thunberg is on board, and Max Factor, Nobel Prize in Economics. Many scientists and artists, a handful of politicians and actors. The Austrian chancellor because they wanted to get rid of him, some European presidents... We have 150 passengers on board."

"Thunberg? She's the one who won the Nobel Peace Prize. She must be a hundred years old."

"She's 87 years old. More precisely, at the age of 87, she was the victim of an attack of the Doomsday adepts. Her family agreed to have her cryogenically preserved for a better future."

"I'd like to talk to her."

"In nine and a half years. She's asleep."

"And the crew members?"

"Military and civilians from the recovery tanks. Strategists, engineers, biologists, medics, scientists. We needed people who had the knowledge and confidence to establish a colony four light-years from Earth. Shortly before launch, the reconstruction technology was fully developed. The problem: you need zero gravity for medical reconstruction. The cryotanks—largely with victims from bomb assaults on military units, also some civilians—were transported to L1 to start the reconstruction. By chance, the spaceships were built there. You spent the last few years in a cryotank between Earth and the Moon.

Then the money ran out, cryopreservation became too expensive, EUROFORCE stopped the medical program. They would have abandoned the tanks.

On the other hand, it turned out to be a good thing. The ships needed a crew. There were over a thousand tanks at the libration point, a few kilometers from the two spaceships. Most of them skilled fighters, many with specialist training, disciplined, pragmatic, decisive. Just the sort of people an expedition to the New World needs. The regenerators were on board anyway. The commanders made a selection."

"And you just"—he waved his arms—"took me along?"

"Like I said, the committee wanted you on board. Your CV seemed to impress not only the captain."

He looked down, trodden to the ground. She assessed him with a mixture of contentment and concern.

"Come along," she said spontaneously. She took him by the hand and pulled him in wide jumps behind her toward the bionic center. The centrifugal force decreased steadily as they crossed the belt and set course for the central igloo. Finally, they almost floated through the door. Inside the igloo, the light was dimmed, and water pumps stamped softly. On the right-hand side, a massive security door blocked the way. Next to it, a dead monitor. The surroundings reminded Storm of an underground laboratory he had visited long ago. Padmé pointed at the door.

"That's the crate, that's where the crews and passengers are."

He approached the door, but she pointed at the shield and pulled him away to the right.

*No entry. Cryostasis* was written on the door.

They floated through an open bulkhead into a hall. Tubs stood close together. Many were empty, some contained a semitransparent liquid. There were cables, metal rods, and pulsing hoses. Storm came closer to one of the tubs. A face covered with blond hair glowed pale through the liquid. A young woman, asleep.

"Is that the—tanks?" he whispered.

"That's right," she whispered back. "This is Nordica. She'll wake up next."

*A blond beauty from the Arctic Ocean*, Storm thought.

"What about her—?"

Padmé pointed to the lower part of the tub. Where the pelvis of the sleepers should be, there were surgical instruments and tiny threads at work, fanning like anemones in the nutrient solution.

"She is being reconstructed. An explosion severely damaged her pelvis. The internal organs are done, so it will be faster from now on."

*I looked like that*, thought Storm. *Helpless and incomplete. They gave me back my life.*

"I suppose I should be grateful," he muttered vaguely.

"You should."

"And you, Padmé? Are you a—a tank woman, too?"

She smiled. "I'm real, so to speak."

| | |
|---|---|
| Distance to destination | 3.038 ly |
| Time to arrival | 8 y 213 d |
| Distance from Earth | 1.205 ly |
| Speed | 0.472 c |
| Acceleration | 0.094 g |
| Slip correction | 76.25 d |

# Episode 8 Training

#news flash 1.7.2215

In the Kabul Treaty, India and Pakistan renounce the use of nuclear weapons. China and Afghanistan guarantee the borders.

Core Europe rejects the Eastern European states' demand to double EU
  funding. Greece, Bulgaria, Romania, and Poland leave the EU.
© RG channel via moon base Clarke II, sent 17.4.2214


Eleven months had passed since crew 7 had been sent back into
hibernation. The officers in charge had changed. The new captain was Sven
Molander, Thelen had been replaced by Nils Kvalheim, and the new
quartermaster's name was Lilly Angelis.

A few days ago they had been released from the crate. Storm was as refreshed
as after a good night's sleep; the only problem was getting used to the artificial
gravity again. The first day was dedicated to acclimatization. He followed
Lilly's tip and jogged first on the small promenade, then on the big one. Also,
he went first in the easy direction, then in the difficult one, just as she had
recommended. (If he ran against the rotation sense of the EXODUS, he could
perform long, liberating jumps, but in the opposite direction his steps were
much shorter. Intuitively he understood the difference; because in one
direction he subtracted speed, in the opposite one he added it to the rotation
of the ship.)

On the promenade Jason Nygard came toward him, jumping like a
kangaroo in the easy direction. Panting, they stopped in front of the arena.

"Long time, no see."

"A year goes by pretty damn fast."

"What happened to the pillory?" Storm asked.

Instead of the neck irons, a hologram was displayed at the capital: The
EXODUS in orbit around her new home.

"Molander wants to motivate us. He loves triumphal columns."

"And Ahlgrim..."

"...rebuilds them into pillars of shame. It's like the wolves and the sheep. It's
an economic cycle. Riccati's equation, no mystery."

They were silent, watching the Belt with the joggers. Nygard shook his
head regretfully.

"What?"

"Ah, nothing. I just reckoned that would be enough...why am I giving
physics lectures and demonstrations?"

"I don't... "

"Clockwise and anticlockwise. It makes all the fucking difference, doesn't
it? You can see that, what am I saying, you can *feel* it in your bones."

"Sure."

"That's what the Coriolis force does. When you run against the rotation, it
points up, when you run with it, it points down."

"Yeah, sure. Against = kangaroo, With = snail. Got it."

"And yet there are people who believe that the ship doesn't rotate."

"Oh, yeah?" Storm was curious, so he listened intently.

It looked like Nygard was wondering if he should say more about it. "Do you feel the wind?"

Storm lifted his head, turned around, took a whiff. A barely noticeable breeze fanned his face.

"Wind I wouldn't call it. A breeze, yes."

"That's because the habitat is spinning. Air vortices, like a miniature low-pressure system. Coriolis forces."

Storm nodded. He knew that the air movement in high- and low-pressure areas was caused by the Earth's rotation, and similar things were happening here.

"A year ago I wasn't sure, I couldn't feel the wind. But the anemometer proved it. Physics is relentless."

He sounded regretful, as if he hoped to prove otherwise. *A strange guy*, Storm thought.

"Come and see my next demo here at the arena," Nygard said.

The inspector held him back. "Listen, I have one more stupid question."

"There are no stupid questions. Only stupid people."

That wasn't very motivating, but Storm asked the question anyway.

"The information there on the display"—he pointed at the 200-inch screen right above them—"ly means light-years. How much is that?"

"You don't know?"

"I just know it's a hell of a lot."

Nygard cleared his throat like he was preparing for a lecture. "A light-year, my dear inspector, is the time that passes when light travels a year," he lectured with a serious face.

"Ah, yes." Storm thought about it. It sounded pointless. "In law school, we probably would have called it a sophism."

"Not bad, Commissioner. That's nonsense. Most people fall for it. It was foolish, I apologize formally!"

Nygard hinted at a bow and continued while they were jogging side by side in long leaps:

"In one second, a flash of light travels some 299,000 kilometers. In one year that makes a distance of 9.46 trillion km. That's one light-year."

Storm tried in vain to imagine that number. He looked up at the display. "Atlantis is a long way off," he said helplessly.

"While we're at it," the physicist continued, "velocity 0.47 c. We're flying toward Proxima at almost half the speed of light. The acceleration is just under a tenth of a g. Up there"—he pointed to the pipeline, where several bivouac-like tents oscillated inertly in the airstream—"there is no centrifugal force, only the thrust of the engine. Up there you weigh only a tenth of your earth weight. Ideal for certain relaxation exercises."

When Storm didn't say anything, he went on. "Those are the jumpers. You can rent them."

Nygard smiled mischievously. The tents were made of dark fabric. Two of them seemed to be occupied.

"Jumpers?"

"They bounce up the pipeline and glide slowly back down again. Until then, you are weightless."

He pointed to the igloo. A jumper just arrived there, lingered a few seconds, then slowly climbed back up.

"Or you can hold it in place. Like that one. You'll only weigh a couple of kilos. That's not bad. You can use your imagination, if you follow me."

Storm found the idea of weightless erotica strange.

"What is slip correction?" he changed the subject.

"Our clocks run slower than those on Earth. We've tuned them so we don't mess with the date. They show Earth time. Just like the clocks onboard the Galileo satellites."

"Is that so?"

Nygard nodded assiduously. "So far, we have won 76 days over the Earthlings, or in other words, you would be about 11 weeks younger than your twin brother who stayed home—I mean, if you had one."

"Why?"

"Well, it's complicated. It took Einstein years to understand it. Say time is flexible."

*How right you are*, Storm thought as his memories flashed up. "Why does it take Earth's news so long?"

"Well, you know, a light-year is the time it takes for light to travel..."

"...if it travels a year," Storm added.

Nygard's giggles turned into a fit of laughter.

***

"Welcome to the VR group. I'm David Müller, your instructor."

"This is Oliver Storm, your future boss."—Müller pointed to Storm—"a fabulous detective. Dana, Katrin, Silke, we've met. I'll get to know the boys too, or rather they'll get to know me. You're new here and I'll introduce you to VR.2 technology. VR.2 is the second release of an immersive virtual reality. You have been chosen for your character and physical ability to form a rapid reaction force under Oliver's leadership."

They had gathered at VR headquarters, the training room. Storm nodded at the group. He had already talked to everyone—Katrin (called Cat), Silke, Dana, Tom, Florian, Bill, and Steve. Athletic young people, self-confident and attentive.

The instructor took a kind of hood from the desk. An elastic anthracite-gray cover made of hexagonal honeycombs.

"This thing"—he stretched it like a bathing cap—"is the latest generation of EUROFORCE's Rift technology. It was originally developed for combat pilots and elite troops to prepare them for their missions. It places the wearer in a virtual training environment. The principle is simple: the extremities are stupid as hell and feel nothing. Everything happens up there."

He pointed to his military haircut.

"To create the impression of movement, all it takes is to get the right neurons in the cerebral cortex to fire. It's the same with the tactile nerves of the skin, the proprioceptors for body sensation, the nociceptors that signal pain—in short, VR.2 couples virtual reality directly to the brain. You know the principle from computer games of your—eh... era. Transcranial neuron stimulation. At that time, magnetic fields were used, the accuracy was miserably poor. They have made progress in the last century."

He proudly tapped the hood and continued: "Terahertz lasers are located in the honeycombs. They emit microwaves that stimulate the afferent nerves through the skull. Terahertz radiation penetrates bone and tissue. The focusing on the neurons that code for movement, orientation, balance, organ feeling, haptics is so precise that virtual space is almost perfect."

He pulled a schematic drawing on the screen. A hexagon of the elastic canopy appeared. Inside it, there were numerous light-colored rings.

"The centerpiece is this quantum-dot laser. These ten transmitters"—he zoomed in on one of the rings—"are driven coherently, so the waves have a fixed phase relationship. We can focus the radiation to a focal spot less than ten micrometers in diameter. Some tricks from the STED technique improve the resolution by a factor of ten."

Storm had read up on this, so he knew what Müller was talking about. VR.2 was a further development of the miserable virtual reality of Underworld, that game from an earlier life. Something about radio waves, the seller had told him.

The Commissioner looked around discreetly. Everyone nodded.

"For EUROFORCE, VR.2 is extremely useful. It saves us a lot of maneuvering. For us, VR.2 is even vital, because we can't train outdoor maneuvers on site. Any ideas why that's not possible?"

"The radiation," explained Florian to the group.

"That's right. We are flying at half light speed toward Proxima Centauri through interstellar gas. We are exposed to the particle stream coming from the bow. It is like a hurricane blowing in our face. Half an hour outside, you'll get five hundred millisieverts. If you practice this repeatedly in real life, say to fix the antenna after a bolide hit, you're dead."

Müller looked over the group as if he wanted to check if the message had sunk in.

After much hesitation, Steve raised his hand. "One question, Sir: Why are we safe on board? At half the speed of light, protons fly through centimeter-thick steel like butter."

"That's right. And we only have thin aluminum and compound walls," Müller confirmed. He raised his hand to his chin and frowned as if he was pondering hard.

*The hydrogen is sucked into the pipeline*, thought Storm.

"The protons are sucked away," Müller said. "The pipeline is a gigantic vacuum cleaner. But the remainder that gets through is still a problem."

He stretched out his hand, spreading his fingers. "They would shoot through my hand like bullets. Only halfway through my body, they would come to an abrupt halt, inducing an avalanche of damage to the inner organs."

He drew a sketch of the EXODUS on the screen, zoomed in on the bow. Like a voracious mouth, the suction port of the pipeline sat in the center. A ring of honeycomb-shaped metal segments covered part of the front of the ship.

"The water tanks," Müller explained, pointing to the honeycombs. "We have about four hundred tons of water on board, which is constantly being recycled. The tanks are 25 centimeters thick, they cover the habitat. The protons come from the front at half the speed of light. They have a kinetic energy of 190 million electron volts but they cannot penetrate this barrier. Water is simply fantastic. Only a few ultra-fast particles from the GCRs can get through."

The listeners remained silent, perhaps disturbed or relieved.

"At the rear, the engine would grill us," one of the women said, expanding upon the subject. "That's right, hundred million degrees in the core, that's a lot of waste heat. We would have to shut down the engine for every training session and wait days while it cooled down. Besides, it's risky. Ignition is a delicate process, anything can happen."

More illustrations followed—canopy layout, communications, bandwidth diagrams, encryption.

"The Rift you've seen in the past creates a fairly passable virtual reality. Some games even allowed several people in the same environment. I don't need to point out the spicy details, I'm sure you chiselers are familiar with it and were disappointed despite the cool stuff.'

He paused in vain, expecting a reaction.

"Yes, it didn't feel real. But VR.2 is different. The software can control ten helmets. That means the whole group is in the same virtual space. You will be able to see and talk to all members of the command. And! you can touch each other, you'll feel everything"—he put a hand on the shoulder of Cat, who was sitting in front of him, stroked her neck and suddenly feigned a strangling movement—" as if it was really happening HERE!" Before she could react, he threw the hood at the inspector and clapped his hands. "Do you want to try this?"

An eight-voice "Yeahh!" was the answer.

Storm carefully took the thing in both hands. It seemed to pulsate, the honeycombs were constantly deforming, as if they were impatiently waiting to unleash little young bees on his brain.

"Don't get the idea of doing anything crooked. You won't see me, but I'll be watching you. Every single one of you!" Müller railed.

Storm put the hood over his head and put the Rift on. An orientation phase, the image flickered, solidified for seconds, as he knew it from the ultra-fast particle showers, then virtual reality built up. Ants tingling on arms and legs, a brief feeling of floating, then all eight participants sat at their desks as before, only the instructor had disappeared. Storm got up, touched the table, looked around, went to the others, patted them on the shoulders, spoke to them. Trivialities that sounded helpless and insecure because VR seemed so real. The faces were free (in VR), although everyone had a Rift before their eyes (in reality). Storm scanned his face, felt cheeks, nose, eyes. *Crazy. Somewhere in my brain, the Rift that I wear disappears. I can't even get out of the simulation*, he thought fleetingly.

Suddenly Müller appeared as if from nowhere. The inspector considered that the session was over, but when he inconspicuously groped for the data glasses, there was nothing but nose, eyes, skin. They were still in VR.2.

"Have I told you about the emergency exit?" Müller asked, knowing full well that this was not the case.

"Let's catch up now. What do you do when you want to get out?"

There were murmurs, quiet discussions. Some reached for the Rift or the hood on their heads—in vain.

"I guess we can't take the helmet off," Katrin said. "It doesn't exist in VR.2."

"We'll tell you over the radio," suggested Florian.

"That's possible. But I am also in VR.2."

"We'll wait for you to bring us back," Storm said.

"That's right. That's the normal case. But you might get into an emergency situation. Maybe you feel really bad—I mean really bad, heart problems, circulation—then you go to the *exit point*, which I'll tell you before every mission. This time it's there!" Müller pointed at the wall mirror. "You go there, then you press firmly against your right temple, like this—" He demonstrated it and went through.

They stared at the wall in amazement. Tom got up to follow him, but then Müller stepped out of the mirror again.

"Wait a minute, you're not getting away from me that easily. We're trying to simulate an emergency." He looked around sternly. "The room here is no different from the VR center. We're still on level 2, gravity half a g, okay?"

They glanced at each other, looked around the room, nodded hesitantly. They waited for further instructions, but Müller kept silent. After a minute Storm felt that the air was getting muggier. He looked around the room. The others also seemed to feel uncomfortable. Then he noticed that his shoulders got heavy.

"Is something wrong?" Müller asked facetiously.

"The centrifugal force," Storm suggested.

"Right so! I increased the rotation speed of the Exodus—only here in VR.2, of course. And it will continue to rise. Your task is to hold out as long as possible."

He stood up, stepped in front of the mirror, waved as if to say goodbye, pressed his right temple, and disappeared. The centrifugal force increased steadily. Storm soon found it hard to breathe. When he got up, he had lead weights on his shoulders. His back was sore, his legs were barely supporting him.

Silke got up too, pulled a face and stomped to the exit point.

"See you soon," she gasped and let go. Storm stood beside the mirror, supporting himself with his hand. The force increased relentlessly.

"Come on!" he cheered his crew on. They came stumbling and panting. Everyone went through the mirror except for Dana, who didn't want to give up yet and was now crawling toward the exit point. Storm wanted to help her, bent down, and collapsed. Steel clamps pushed him to the ground, he couldn't breathe anymore. His field of vision narrowed to a tunnel in which everything was spinning. At the end of the tunnel lay Dana, very small, far away. A hand stretched out, creeping toward him like a snake. He pushed his right arm toward her, it turned with the tunnel. *Now it's going to break off*, he thought as everything went black.

Suddenly he was no longer forged to the ground but sat upright. His breathing calmed down, the pain in his chest and back subsided. He took off his helmet and Rift. There he was again—in the VR center with the others who seemed more affected than exhausted. Dana sat leaning forward, not moving.

David Müller gently took off her helmet and massaged her shoulders. She moaned as she emerged from the simulated blackout. Dazed, she looked around and smiled at her future boss.

"Did it work?" she croaked.

"Yes, it did," Müller confirmed. And to Storm, who stared at her uncomprehendingly: "She got you out before she collapsed."

During the next week the instructor introduced them to the details, and little by little they became more confident in dealing with VR.2. The Commissioner carefully built up his troops, observed their behavior in simulated decision-making situations, their ability to communicate, their contacts with each other. Everybody had to master everything, but there were specialists—attack, strategy, overview, defense. Physical training took place in the Belt. Here, the rotation speed of the EXODUS could also be changed in order to accustom it to different scenarios. They took Jason with them as an observer, and he commented on the mechanics of their exercises. With time they mastered the unusual movements under the influence of the Coriolis force. They learned how to throw objects at each other over long distances, and how to use firearms.

The near-reality of VR.2 impressed Jason.

"And if we put helmets on here and teleported to VR.3... as realistically as here, how can we know what reality level we're on?" he mused.

At the end of the month, they reviewed their training goals in real life. Storm was skeptical at first, but when his troops on the promenade easily mastered the most difficult exercises they had practiced for so long, he was convinced.

Now they were ready for the real thing. Müller presented him with the scenario: The mission was an armed confrontation; the commander had to be protected from mutineers and the uprising had to be crushed. Storm was surprised, because such a situation was highly unrealistic.

"Not so much," was Müller's answer. "There are some in the crew who are not happy with the captain."

The instructor explained to him that they had to use the weapons in case of danger to the captain's life, to carry out the commander's orders, and in case of hindrance those of the first officer. Storm wanted to know who they were supposed to be firing at.

"You'll be shooting at softies."

"Who?"

"Softies are software-generated avatars. There are no players behind them."

"What if they shoot at us?"

"Well, that would be bad. Your avatars are hardies."

"Eh?"

"Real players, human hardware, so to speak. You may have bulletproof vests, but you can still get hurt. And if your people are mortally wounded, they're dead—only in VR.2, of course—but unfortunately, there's no resurrection. Nobody can get you out except me."

The exercise was successful. The mutineers were overwhelmed. Molander ordered a summary trial, the three leaders were tried for mutiny and high treason and executed. Unfortunately, there was an accident: Silke's avatar was mortally wounded.

"You were really dead, you couldn't be any deader than the way you lay there," Florian commented. "How did it feel?"

"It was spooky. I felt a violent blow on the head, everything turned, then I—floated away. I saw myself lying there, from above. The assassins faded, somehow they became transparent. I don't know—. Then there was only noise in the Rift, I felt that I was sitting there and took off my glasses and helmet."

A discussion about the appropriateness of Molander's virtual decision came to nothing. Legally, the verdict was unimpeachable; threat to the mission by conspiracy theorists, martial law, unrestricted command. *And yet, it wasn't fair*, Storm thought. He couldn't get the exercise out of his head for days. It had been so realistic; at times he had forgotten that it was a VR.2 scenario. He felt sorry for the three convicted softies, as crazy as it was.

The training continued. David Müller looked exhausted, and sometimes he nodded off during the debriefings. The inspector quickly took the lead; the most important thing was the cooperation between his folks. He was always planning new scenarios—accidents in the Belt, failures in the control units, bolide defense, vacuum leaks, emergency power supply. Then they moved on to field operations—at the Spider, the landing craft, the Pipeline, the lotus leaves, the engine. Everything was played out until every mission went smoothly.

At the end of their monthly shift, not only David Müller was exhausted. Storm didn't mind getting into the crate. He even looked forward to the long sleep.

| | |
|---|---|
| Distance to destination | 2.535 ly |
| Time to arrival | 7 y 213 d |
| Distance from Earth | 1.708 ly |
| Speed | 0.531 c |
| Acceleration | 0.083 g |
| Slip correction | 125.9 d |

# Episode 9 Rendezvous

#news flash 1.7.2216
Resistant super germs cause over a hundred million deaths in Europe and the USA.
Russia and Poland conclude a treaty of friendship. Russian troops are stationed at the Polish–European border.
© RG channel via moonbase Clarke II, sent 15.10.2214

No sooner had they started the new shift than they were ordered into the arena. The hologram above the column had disappeared, neck irons decorated the pillory again. Triumph had turned back to shame. Storm listened, dazed, as Thelen announced that they would soon cross the path of the EXODUS I. The fragments of the exploded ship had formed a spreading scattering cone which they would fly through. The hydrogen from the destroyed tanks had already been detected, and the magnetic funnel of their ship would pick it up during the short passage to increase thrust. The risk of collision with a fragment was low, but they would have to be on their guard. At a relative speed of a good 60,000 km/second, a fist-sized fragment would puncture the hull like paper. A hit on the meter-thick radiation shield would release the energy of several Hiroshima bombs and prematurely end their mission lethally. An immediately scheduled mission briefing went to Storm's satisfaction— they were well prepared by the VR.2 exercises, and everyone knew what to do in case of an emergency.

Two days later the time had come. The images from the deep-space telescope were transmitted to the large monitor on the bridge. The debris of EXODUS I was visible in the reflected starlight as a barely noticeable spot directly in front of them.

"Request trajectory," Storm demanded, and a green line appeared that just missed the spot.

"We will pass the scattering area at a distance of 10,000 kilometers, " Müller commented. "No danger from the big lumps. But our telescope unfortunately cannot see the smaller troublemakers. Some might have drifted further away."

"What about the hydrogen? Can it be made visible? Then we would have an idea of the size of the dispersal cone."

"Good idea," mumbled Jason. "Overlay Lyman and Balmer filters, that should work." He tinkered with the telescope controls. The image on the screen changed. The stray light was replaced by a diffuse cloud that filled half the screen. "We're going to fly right through that cloud. Good for our afterburners."

"Can you tell us how far smaller chunks have diffused with the cloud?"

Jason pulled a face. "Am I God?"

"Well, perhaps you will bestow on us your wisdom and expertise. An appraisal would be fantastic."

"An estimate is only worth something if you know the confidence intervals."

"Uh-huh." Storm considered what the better alternative would be: studying mathematics to understand Jason or replacing him with a normal person.

"No problem." David came to the rescue. "Jason, we love you, but just give us a non-binding estimate of the risk of collision."

"Hmm, noncommittal? You sure?"

"Yes, guaranteed!" He looked around. Everyone nodded eagerly. Even the reserved Thelen moved his head after seconds of reflection, which could be interpreted as approval.

"You see, we guarantee. No handcuffs for an incorrect answer."

"Okay, all right, I would say—but I'm not to blame, if it's different, is everybody clear on that?—would say that—well, ten percent. Non-binding, of course."

"And what does that mean?" Storm asked.

"Which means, damn it, nine times out of ten everything could be fine."

"Of what cases?"

The physicist rolled his eyes in desperation.

"If we were to chase ten ships through there, one would probably have bad luck," he explained patiently.

*But we don't have ten*, Storm thought. *Statistics is nonsense.*

"How much time do we have left?"

Jason was questioning his mablet.

"We will cross the foothills of the litter cone in 36 minutes."

The Commissioner had heard enough. "Prepare for depressurization."

Bill alerted the crew. The oxygen apparatus was put on. Storm sent Tom, Cat, and Steve to the defensive lasers. The three of them were very adept with the software during the exercises. As they approached, the radar would give them one or two seconds to vaporize small debris. He himself boarded one of the longboats with Silke. Florian and Dana got the second boat ready for take-off. They could immediately disengage in an emergency. The picture on their onboard monitor looked harmless. At the edge of the screen, the time until the passage of the stray cone was counted down. At minus 10 seconds something flashed up—the lasers had vaporized a potential impactor. Storm was counting in his thoughts. The next few seconds would decide whether they survived.

Suddenly a bright sound, as if a vesper bell was striking. Then the EXODUS shook under a hit. Storm listened with bated breath as the bending vibration of the pipeline slowly subsided.

"Damage reports?" he demanded when nothing moved.

"Thelen here from the bridge. We got a hit on heat sink 13-7. Here's the picture."

The onboard monitor showed one of the lotus leaves. The camera zoomed in, centered on a dark spot, moved even closer. A hole with jagged, melted edges in the heat sink, where the bolide had hit. Storm took a close look at it, discussed it with his copilot. She shook her head.

Nygard said, "It's small enough, you can leave it like that."

The inspector felt the same way.

"If it had hit the radiation shield, we'd be dead."

"There's something else," the bridge reported after a few minutes. The bow camera scanned the antenna and zoomed in on the image. Residual light amplification made the silver wires shine in the faint starlight like spider threads in a gigantic web. Two of them were torn, and the net bulged up around the defect.

"This thing is bending, the phase shifters are broken, we're losing transmission power," Nygard noted. "You have to fix this. You will be working far outside the strong magnetic field. From there you can just about see into the edge of the funnel."

"And?"

"Take a good look, see if the plasma glows there."

"You're kidding? We're in a high-vacuum."

"Not quite. The supra-coils compress the hydrogen in the funnel. I want to see this. Make a short video."

Storm nodded.

The thruster was deactivated, the rotation shut down. The field operation should follow the protocol of a VR.2 exercise. It was the first time they saw the ship from the outside. It seemed to float motionless in front of the starry sky, although it was racing toward its destination at more than half the speed of light.

The light from the stars and the Milky Way was strong enough to illuminate the outline of the ship. Carefully, very slowly, the longboats approached the antenna. Looking out of the window, Storm saw the glowing exhaust plasma of his longboat's thruster when he opened the valve. The beam formed a distorted spiral as if the valve was not functioning correctly. This was the

Lorentz force in the strong magnetic field of the supra-coils he knew from theory training.

Positioned ahead of the starship's bow, they looked back into the funnel which sucked the ionized hydrogen into the pipeline with the help of strong magnetic fields. They drifted over the spider web of the antenna, a 500-meter-wide veil. The fine metal wires glittered in the starlight. High above the spider, the secondary electron stripper—a carpet of electrically charged carbon nanotubes that ionized the small remaining amount of neutral hydrogen so it could be captured magnetically—fluttered like a circus tent in the wind. But it wasn't the wind, it was the stripper's filigree frame, which, without the stabilizing centrifugal force, performed a slow torsional oscillation. Autonomous crawlers swarmed like tiny bugs around the hole in the film that the impact had torn. The mini robots were constantly occupied with repairing small defects in the stripper foil. Now they busily knotted nanowires into a wafer-thin fabric to close the hole. The sight of this mighty construction was breathtaking. At this tremendous scale, humans were tiny insects on an interstellar blossom.

Before the Mayflower anchored in front of the hole that the collision had torn into the net, Storm saved a short video of the funnel with his helmet camera for Jason. There was no plasma glow visible to the naked eye.

The gripper arms of the shuttles were too rough for the repair. Tom was already in the loading bay for the extravehicular operation. He clipped finger-thick light metal tubes together to the required length and approached the perforation. Using a circular saw, he straightened the edges where the melted spider threads formed thick solidified drops, clipped the spare parts to the ends, and activated the phase shifters. After twenty minutes the hole was mended. Storm looked at the spot with the zoom, checked the receiver signal, and gave his okay to return. Meanwhile, the crawlers had fixed the axial stripper. Far ahead flew the armada of main strippers, tiny rings of monoatomic graphene glowing in the glare of Alpha Centauri. Infrared lasers on the battle station of EXODUS accelerated the ultra-light disks continuously by targeted shots to stabilize the armada in front of the magnetic funnel.

The Commissioner knew the Centaurus constellation. He could even visualize the connecting lines between the constellation's stars, where the ancient Greek had imprinted a mythical personality, half-horse, half-human. There was Alpha, so bright that it hurt the eye, just a few degrees from their still invisible target Proxima. In his memory, the constellation was wider; especially the southern cross, between the horse's legs, appeared ridiculously compressed. Certainly a fancy physical effect. *I should have attended Jason's lectures on astronomy*, he realized, conscience-stricken.

The communication center reported the correct function of the directional antenna a little later. After less than an hour they were back on board. The dosimeters showed a dose of less than 400 milliSievert. Tom had received 540 from the long extravehicular mission.

In the evening they hung out at Coriolis Dance to celebrate the success of the mission. Jason wanted to know what the inspector had seen outside. Storm showed him the video and explained repeatedly that there was no glowing plasma.

"But it should be there," Jason murmured thoughtfully. Then he shook his head, as if to get rid of a bad aura. "Never mind. Let's celebrate!"

The training left enough possibilities for leisure. Storm joined a handball team, not so much because he liked the sport but because it made it easy to sweep memories of the past under the carpet. A past that had happened to a person that vaguely resembled him. It was over. So what?

He profited from the VR.2 training. The game with balls that followed exotic tracks was crazy, but the team was not his cup of tea. Too little strategy and cooperation. He switched to a group that did 3D gymnastics. It was fascinating to watch the experienced players as they flew through the belt on precise Coriolis tracks, performing somersaults, pirouettes, screws, and never before seen figures with playful ease.

Storm experimented with augmented reality sex, attended meetings that would have been called swinger parties in another life, and tried to land Padmé, but she kindly and firmly refused. The highlight of a brief liaison with a woman his age whom he had met at a sex party was microgravity sex in the Jumper. The bivouac gliding along the pipeline was made of delicate fabric, translucent on the inside and opaque on the outside. The climax was less about the partner than about microgravity. His experiences were on the whole satisfying, but emotionally frustrating. And something must have gone wrong during the reconstruction, because he had no ejaculation despite orgasms.

The sensation of taste and the foreign feeling on his skin improved slowly. His nightmare continued to plague him, though rarely. And sometimes the events seemed to be frozen for seconds like a stopped video, and he saw himself from the perspective of a drone hovering above him.

After a psychological test that confirmed Storm's stability and resilience, de Vries called him in and handed a document over to the patient. Written on old-fashioned gilt-edged paper, the Commissioner was declared *faithful pilgrim*. The document was signed by Ahlgrim and stamped by the High Command of the Space Forces.

"Wow! The permit to enter the New World?" Storm shook his head skeptically.

"Ah, forget it. That's psychological baloney."

They became friends. De Vries was not deterred by Storm's sarcasm. They had a professional understanding with each other; they often chatted about modern forensic methods. Storm realized that although the hundred and thirty-seven years he had overslept had brought a revolution in stem-cell-based reconstruction, medical forensics offered little that was new. De Vries also enlightened him about some of the details that puzzled the inspector: Out-of-body perception came from the high-energy helium atoms in cosmic rays. As neutral particles, they could not be deflected by the magnetic field of EXODUS and sometimes penetrated the protective water shield. They caused lesions in the body; impacts on the retina produced lightning, and when certain neurons in the brain were hit, the perception of time was confused. Déja-vu experiences, film tears, or out-of-body perception were the result. The intense cosmic rays were also responsible for the lack of pregnancies on the EXODUS; too high a risk for fetuses, and therefore, a small genetic modification had been made in the reconstruction of the male crew—they had no sperm. The few that were not reconstructed were vasectomized. This presented no problem for the planned colony on Atlantis, explained the doctor—there was a carefully assembled egg and sperm bank on board.

He got on well with Jason Nygard. They often went for a bland beer at the ship's canteen, where Storm listened to the story of Nygard's life without the desire for revealing his past. Over time, it dwindled, and it felt good. Jason had studied physics at the Massachusetts Institute of Technology, then fled to Europe when the EPA—the Enforced Patriot Act—made unauthorized departure from the USA a punishable offence. He spent some time working on guided weapons systems in Sweden and finally accepted a lucrative offer from EUROFORCE to train as an officer. Stationed in Toulouse, he had been piloting drones against Arab militias when a bomb attack reduced the control center to rubble.

"I've been here since the beginning. Too long. I ask myself more and more often what I'm actually doing." He made this confession to his fifth beer, which nodded in wisdom.

"You're a hero," countered Storm. "Your children—I mean the new generations on Atlantis—will tell fabulous stories about you. Hey, dude, you're on your way to a new world!"

"If it's true."

"Sure it's true. We'll be there in seven years!"

"If it wasn't for the C-14."

"C-what?"

"I measured something and I don't understand it."

"What? You think something's happening? Technically, everything's under control, right?"

"No, it's okay," Jason reassured Storm. But he smiled mysteriously, as if he knew better.

A week later a new crew member emerged from the tank. The blonde beauty from the Arctic Ocean. Storm was there at her welcoming ceremony in the arena. Ahlgrim again performed his jump from the terrace, gave a similar speech as he did for the Commissioner. Nordica Henderson was a tall, middle-aged woman. Ice-blue eyes, an alert look, a face that radiated determination and self-confidence, the blonde hair tied back into a ponytail. She was the victim of an assassination attempt. This time no unsavory details of her injuries were revealed. After the official part of the ceremony, Ahlgrim pressed close to Henderson and flirted with her, giving her the-big-man-of-the-world routine. He didn't let anyone else in, elbowed her into a quiet corner, and whispered something into her ear, whereupon she smiled politely and took a step back. *What a stupid pick-up line*, thought Storm.

After the commander had withdrawn, visibly frustrated, the inspector offered her his help, since he, as a newcomer, knew the acclimatization phase well. She nodded politely, chatting away before Padmé took her by the hand like a disarranged child. She was exhausted and made a perplexed, confused impression. *I must have seemed that way*, Storm realized.

Over the next few days, they met again and again. She exerted an attraction on Storm that he hadn't felt in a long time. Her face looked familiar to him, or was it her movements that reminded him of Carol? Storm explained things to her that seemed important to him, but Padmé had of course already done that. He felt useless and pompous. In sleepless nights, he felt infinitely alone. He was an island, like in one of those old songs from a century long gone. They were all islands, and often he thought he had to make more of an effort to make contacts. But in those nights he realized with extreme clarity that this would lead to nowhere, must lead to nowhere, because every single human island was in its own universe. That was the outrageous, the incomprehensible, the certain, perhaps the only certain existential truth. And VR.2, that incredibly real second existence, was a shameless extension of being alone.

Padmé noticed his depression and advised him to talk to de Vries or at least consult a PsyBot. But he did not want to. It seemed absurd to him to take therapeutic action against a fundamental insight. He pondered, groping his way from the outrageous human loneliness to the fact that he was in a tiny habitat of hope, protected only by a fragile shell from the infinite emptiness out there, from which it followed that human ingenuity did, after all, bring about something that, via a bleak individual fate, pointed to something greater—namely, to send the desolate islands they all represented on an interstellar journey into a shared future. And if they ever landed on Atlantis against all probability, yes, then—nothing would have changed in the fundamental realization of being alone. And so he tormented himself for many days, his thoughts went round in circles, he felt like a madman who'd lost his way in a blizzard.

In the end, the doctor gave him a drug that was supposed to raise his serotonin level and actually did. His fundamental insight still seemed fundamental to him, but it no longer prevented him from being interested in his environment. Above all, he wanted to know more about Nordica Henderson but did not quite know how to begin. It seemed unseemly to him to ask about things that had destroyed her existence over a century ago. So one day, he read her personnel file, and he was flabbergasted.

*Nordica Henderson, 38*, it said, *living in Vienna, head of the development department at THZ, a Belgian high-tech company for terahertz wave–brain interfaces. Unmarried, childless, seriously injured by a bomb explosion in close proximity to the body. Foreign DNA at the crime scene assigned to a Carol Beauclere (missing), investigation discontinued.*

The attack on her came five months after the strike on Storm. Nordica Henderson was the seventh victim in his well-known series. And the inspector knew who was behind it. The shock of disillusionment replaced the first elation when he realized that this was a cold case—one hundred and thirty-seven years and 20 trillion kilometers were a guillotine that divided worlds.

The next day he invited her for a drink and immediately spoke to her about the bombing.

"I thought you might contact me," she said. "I read your file. Strange coincidence."

Storm sipped his whiskey, which didn't agree with him. He was still suffering from altered taste sensations. "I don't want to offend you, but I'd like to know more about the attack. Do you have any idea who did this to you? And why? Do you remember any details?"

He had to find out Zigmund's motive. Had she been in contact with the RAVEs? Or maybe she knew something about Roland Petrides, the second victim?

She tilted her head back and closed her ice-blue eyes.

"I do not know who or why. I had no enemies, only envious work colleagues. A great post it was. We had success with terahertz technology, and by cooperating with RealGames we were almost as far along with our virtual reality as we are here." She shook her head in amazement.

"What...?"

"It's odd. The VR.2 glasses you use for the exercises."

"What about them?"

"They look almost exactly like our prototypes back then. On the right temple was our company logo. Now there's an aluminum strip right there."

"They have taken over the model. Perhaps your company no longer exists."

"The same design? A hundred years later?" She smiled skeptically.

Storm wanted to follow up on this, but before he could think of a question, she went on:

"Of course there's competition. If you're not envious, you're out of your league. But it was all in the usual professional setting. It doesn't justify a bombing."

"Does Roland Petrides mean anything to you?"

She thought about it, then shook her head. "In any case, nobody I know would have had the opportunity to arrange something like this."

"Why?"

"I was just out jogging. It was late at night. Suddenly, I hear the sound of engines. A drone lands less than 10 meters away. A EUROFORCE Medibot."

A picture from his nightmare forced itself upon Storm. The Kraken with the sign of Asclepius. Had there been a EUROFORCE logo?

"He's coming closer," she continued. "He wants to give me a shot. High blood pressure, he pretended. I know my blood pressure. When I go jogging, I wear an extra biosensor. I knew something was wrong right away. I always carry a gun, you know it's dangerous on the Wien River at night. I tell him to stop, but he comes closer and I shoot his camera. I'm sure I hit it. Then there's only white noise."

"The bomb. Did you see it?"

She raised her shoulders, played with her drinking glass. *Slim fingers you have*, Storm thought.

"The Medibots have a kind of tentacle," he explained. "With servomotors at the ends."

Again she closed her eyes. "There was something, perhaps. A stick, a short stick."

"A cylinder?"

"Could be."

"Color?"

"Red, maybe. Red and black."

Storm watched her sipping her whiskey. The ice cubes clinked in the glass. While she was talking, his unconscious had become active. She had mentioned RealGames and he remembered an amused smile. A smile from the distant past. He decided to push forward.

"Could it be that we have met before? " he asked.

"Where would that have been?"

"RealGames, the start-up with the VR software. Have you ever visited it?"

"Sure, several times. We had a collaboration with that company."

"Is it possible that we met there?"

She frowned but said nothing.

"We crossed each other's way in the foyer. Apparently, I amused you."

She stared at him, eyes wide open. "The man with the smiley face bag? Was that *you*?"

She laughed in surprise, looking at Storm as if she had discovered him anew. Then she emptied her glass and reordered. Fleetingly she touched his hand, still playing with the glass.

"That was so cute!"

| | |
|---|---|
| Distance to destination | 1.985 ly |
| Time to arrival | 6 y 213 d |
| Distance from Earth | 2.258 ly |
| Speed | 0.552 c |
| Acceleration | −0.098 g |
| Slip correction | 186.2 d |

## Episode 10 Thrust Reversal

#news flash 1.7.2217

Iranian militias invade Saudi Arabia via Iraq and the Persian Gulf and advance toward Mecca.

Asteroid Apophis II has been upgraded from Class 4 to Class 7 on the Turin scale. An impact of the 10-km-large body would result in a global catastrophe. The international CATERPILLAR project that aims at diverting the asteroid from its orbit by a giant thruster has been temporarily halted due to the war situation between several member states. Project Mountain Fortress to create shelter in high altitude areas was launched.

© RG channel via moonbase Clarke II, sent 18.3.2215

Crew 7 was back on duty. They had covered half the distance to their destination—two light-years—eight months ago while still sleeping. Crew 11 had switched off the supra-coils and the fusion reactor, turned the ship around so that the bow was facing Earth, and reactivated the magnetic funnel. At this point, they raced toward their destination at half the speed of light, stern first. From then on, the ship slowed down, because the widely spread magnetic field of the funnel acted like a brake sail in the wind of interstellar hydrogen. The autopilot controlled the supercurrent just to slow them down by a tenth of a g. If necessary, the reactor contributed counter-thrust from the hydrogen tanks. The radiation levels, which had risen dangerously until the braking maneuver, now dropped again, not only because they were traveling more slowly through the interstellar medium, but also because the reactor screen turned in the direction of flight kept the particle wind away from the habitat.

It was a strange feeling to know Earth was off the bow. The nose camera was centered on the sun. Next to it was the "W" of Cassiopeia. Storm searched in vain for the Big Dipper. Surprisingly few stars filled the field of view of the wide-angle camera. The rear camera, however, showed a dense collection of stars. In the center, Alpha Centauri, brighter than Venus in old Earth's sky. Proxima, the target star, could not be seen with the naked eye—without light amplification, it was too faint at magnitude 7.

Jason explained in his demos what they saw. The EXODUS coasted at 56% of the speed of light. Just as rain seemed to fall on a fast-moving car from the front, the light coming from the stars was apparently tilted in the direction of flight—the stars at the apex, as he called their target direction, seemed closer together, whereas in the opposite direction they appeared half as dense. Cassiopeia was magnified twice; it almost filled the field of view of the 110-degree camera.

The Doppler effect was also explained: Previously, they had looked in the direction of flight, and the light from the stars had arrived with higher frequency and energy, which was noticed as a blue shift. All stars seemed to shine brighter and bluer. Now, since they were braking, the bow screen was looking in the opposite direction, the starlight arrived with less energy, and the celestial objects appeared weaker and redder.

"If we look at the whole sky, we could see a rainbow of stars, from violet to blue, green, yellow to deep red?" someone curiously conjectured. Jason waved. This was a popular science-fiction theme, but unfortunately incorrect. Even sci-fi greats like Larry Niven were wrong. The Doppler effect did not make a rainbow in the sky, Jason explained, it only seemed to change the temperature of the stars. Those in the direction of flight appeared hotter and brighter, those in the opposite direction cooler and fainter. Therefore, the sun now resembled Antares, the orange-red main star in the Scorpius constellation. The monitor showed everything in true colors. It was not spectacular.

In the first week of Crew 7's shift, there was a strange incident that turned out to be significant. The commander ordered Storm to the bridge. There would be work for an investigator. Would the exercises with the task force leave spare time? Storm wondered about the captain's businesslike politeness. He did not know him like that.

The incident the inspector was to investigate was a stupid vandalism. Someone had smashed Nygard's gyroscope. They met in the arena. Sitting in the first row of seats, the physicist rummaged in a box and pulled out a wheel with thin metal spokes.

"Do you ride a bike?" Storm asked, amused.

Nygard growled angrily. Storm vaguely remembered school experiments with a similar instrument, but he couldn't recall exactly what he'd used it for. Nygard put the thing down in the middle of the arena. Some red spots were visible on the ground.

"There it was."

The Commissioner picked it up and looked at it from all sides. The spokes were bent, the rim was off-center. Some spokes were missing. The gyroscope was found one morning in the amphitheater in this condition. In the photos he'd requested it was lying in the middle of the arena, as if on display. Someone had sprayed the words "Queen of Hearts" in red paint on the floor next to it.

Nygard took the wheel from his hand.

"Foucault's pendulum," he said.

Storm shrugged, at a loss.

"Doesn't ring a bell, Inspector?"

"Hmm—a novel? Something classic..."

Nygard rolled his eyes in despair and explained why he needed the gyroscope. As soon as the flywheel rotated, the axis of the gyroscope was stabilized. It always pointed in the same direction. If it was carefully positioned near the pipeline, it would slowly drift toward the igloo. Meanwhile, the EXODUS rotated clockwise around the ship's axis, which the spectators didn't notice, of course, because they rotated with it. To them, the axis of the gyroscope seemed to turn just as slowly in the opposite direction.

"This proves that the ship is turning. The force keeping us on the ground is not some artificial gravity, but centrifugal force."

"Sounds logical. But who wants to know?"

Jason Nygard nodded calmly. "Which brings us to the heart of the problem. Actually, nobody here."

"??"

"Most people understand intuitively, and there are quite a few who know the equations. Fighter pilots, some of whom we've reconstructed here, know the centrifugal force firsthand. But there are CE-4s in the crew. I do demos for them. Only—they don't want to know, for them this is"—he made a sweeping gesture—"a deception. So nobody really is interested."

"What are CE-4s?"

Nygard leaned over to Storm and whispered, "Watch out for the aliens! We have to keep it down. They're listening. CE-4—Close Encounters of the Fourth Kind. They believe they were abducted, manipulated, controlled by aliens, they had chips or antennas implanted in them, I don't know. They hear voices, they see themselves lying on operating tables from outside their bodies, they are being experimented on. Mostly on their genitals."

"What's that got to do with the gyroscope?"

"It started a year ago. Probably earlier, but it was done in secret. A small group approached Ahlgrim to ask him to do something about the aliens. They were convinced the crew had been abducted. That the EXODUS was a prison inside an alien ship, using technology far superior to ours to create artificial gravity in a hollow world. But they make mistakes, the aliens. The strangely curved trajectories when you drop objects, the strange feeling when you walk, the change in weight depending on the direction you move. The voices in your head, the GCR blackouts, all that stuff, you name it. Ahlgrim's response was predictable: he forbid them from spreading the bullshit and ordered me to dispel their fantasies by teaching them some basic physics. So once a month I give an introductory lecture in mechanics with some nice experiments on inertia and angular momentum. For that, I need the gyroscope."

He banged his foot listlessly against the deformed rim.

"I would show two or three experiments. That would be enough to convince most of the people. The coolest thing is when I show the Coriolis force..."

He thought it over, looked around, checked the promenade, seemed satisfied with the results.

"Let me show you."

He took a tennis ball out of the box and said, "We need the Little promenade."

They crossed the Belt. Nygard positioned himself where the Little promenade offered free space, checked once more if there was anybody around, no one could be seen overhead. He positioned himself like a baseball pitcher and threw the ball just above the ground against the rotation of the EXODUS, diagonally forward toward the arena. The ball hissed away like a frisbee, climbed up, followed the curve of the promenade, got lost in the ascent high above the pipeline, and came back from the other side like a boomerang. Jason had turned around in anticipation of the approaching ball and casually caught it with one hand. He smiled proudly.

Storm nodded, impressed. "Not bad. Can I have a go?"

Nygard handed him the ball.

"You have to throw it as fast as the EXODUS spins, but in the opposite direction. If you were to look at it from the outside, the ball would be stationary while the habitat kept spinning. You can then catch it again after a full spin. And about 45 degrees forward toward the Great Promenade to compensate for our linear acceleration."

Storm stepped into position and cocked his arm. In the process, his elbow smacked into Nygard's ribs. "Oops! Sorry!"

Frowning, Nygard rubbed his chest. "I guess you never played baseball."

"Well, handball, at school."

"I don't know what to do with you Europeans." Nygard demonstrated the proper throwing movement again. "Don't forget to aim slightly forward and up...to compensate for the thrust of the Ramjet and the air drag."

Storm threw with full momentum. The ball rose high, much too high, and instead of coming back to the ground, it described an upside-down throwing parabola, seemed to stand still above their heads at a height of ten or twenty meters and came back from the other side as if in slow motion. Storm realized that it was going to land far away, so he ran toward it, but the ball, out of sheer malice, followed an increasingly curved path. Storm jumped up to catch it in flight and found that he too was following an impossibly curved trajectory. He missed the ball and somersaulted as he fell. Nygard laughed loudly and helped him up.

"Self-experiment with illusory forces," he commented.

Storm was surprised that nothing hurt him.

"That was cool."

"Hey, you were doing okay. Good body feeling. Maybe you should try knocking on the gym door."

"You should have warned me. How long did you practice?"

"For a week or two, an hour every day. That's enough if you have any prior experience. Before I escaped to Europe, I was a pitcher on the MIT baseball team."

"You are a real friend. And with this lousy trick you convince them?"

"No. The point is you can't convince conspiracy theorists. I've tried to make that clear to Ahlgrim. It's hopeless. He's a stubborn bastard. Molander's better. He listens."

"How many CE-4s are there?"

"Officially five."

"Only five? Is *that* a problem?

"We don't know the real number. Who knows how many people believe that nonsense, but don't dare talk about it. Because they think the aliens are listening. Besides, the captain has forbidden any mention of it. You have to act as if you don't know, you have to lull the aliens in safety. And unfortunately, conspiracy theories have a tendency to spread, as we know from our time."

"You mean the moon landing?"

"The moon landing, the flat earth, chemtrails, the Elders of Zion. It's a wide field."

"Who can believe such nonsense?"

"Well, many. People regularly confuse the strength of feeling with the strength of argument. And as you know, conspiracy theories are built such that you can't disprove them. People believe that if you can't disprove something, it's true. The opposite is the case."

"If you say so..."

"I maintain that Huitzilopochtli created all the world's crude oil. You can argue against it, but you can't disprove it. The claim is scientifically worthless."

"Why would the CE-4 smash your gyroscope?"

Nygard leaned forward to Storm and whispered, "Because it's an alien trick. There's a mechanism built into it to fool us. All fake!"

"Did they find the missing spokes?"

"No. The guy took them."

The Commissioner investigated systematically. The surveillance camera outside the arena was broken the night in question; it wasn't fixed until the next day by the service bot. He went through the personnel files of the suspected CE-4s. All of them had a severe trauma before the hibernation—torture, killed comrades, plane crash, explosions. All suffered from post-traumatic stress disorder. When questioned, they admitted their mistake. They praised Nygard's lectures, claimed to hear no more voices, that they were cured. "Queen of Hearts" meant nothing to them. Four were eliminated because they had been tracked during the critical period. The fifth, who was in charge of the air conditioning in the habitat, had no alibi because he was off duty that day, which meant that his tracking device had been inactive. He asserted to have been in his apartment. What he did there was nobody's damn business, that's why they called it a time-out.

The red paint wasn't much to go on either. All but the supect had spray paint in different colors. Storm felt like an idiot when he was patiently told that they were for artistic recreation. Yes, the art wall. He should have thought of that.

He questioned Jason Nygard again, hoping to find another motive. And he had the vague feeling that something was troubling the physicist. An after-hours meeting was the obvious choice. Over a drink, he asked what on earth one could do with bicycle spokes. Nygard shrugged. Nothing sensible, he said, but one doesn't know what makes CE-4s tick. Radiation protection maybe, defensive antennas—early on, these people had put on aluminum hats so as not to be influenced by aliens.

"Could they check something with this?" he asked, following an intuition.

Jason looked up in surprise.

"Check?"

"I'm thinking of dowsing rods. Maybe they're using the spokes to search for the aliens' radiation or hidden signals from Earth."

Nygaard was sipping his whiskey. "Why not? You have to question everything. So they might think we're still close to Earth. We're not actually on this long trip to Proxima Centauri."

"I wouldn't mind. I'd love it if we could just go out there"—Storm pointed to the rear monitor—"turn right and finally get a drinkable whiskey at my bar around the corner."

Nygard swayed his head, smiling. He looked thoughtful. *He's trying to tell me something*, Storm thought. After two more drinks, the physicist thawed enough to say what he was thinking.

"Check up, you said. That's clever. We do it all the time. Everyone does it. We check on the cell phone, make sure the house is secure. Did the app do my financial stuff? You check if there's enough whiskey in the bar." Nygard was playing with his glass. The ice cubes clinked.

"You check your hair before you meet your date."

Storm smiled back at him. "And your wife checks if there's foreign perfume on your shirt."

Nygard grinned. "It's no different in physics. But we work more precisely. How many micrograms of perfume? What's the composition?"

"Right. You need numbers, just like in forensics. But there's a lot more math in physics," Storm pushed.

Jason nodded deeply, sipped his glass, started to comment, then fell silent.

"What would you do without a calculator?" Storm asked. "You can't check anything."

Nygard shook his head. "It's often easy. You can calculate the thrust of this monster"—he gestured at their surroundings—"without paper. Force is mass times acceleration. 1500 tons times zero point one g is about one and a half million newtons, which is quite modest. The ancient Saturn V rocket's thrust was twenty times greater. Then you calculate how much hydrogen the fusion drive needs for this. All you need now is a piece of paper. Loss of mass during the fusion of four protons to a helium nucleus, $E = m\,c^2$, waste heat, other losses, … that makes one to two kilograms of hydrogen per second, not even an exhilarating amount. Hydrogen is everywhere in space."

"Sounds simple."

"It is. Physics is the science of the simple processes in nature. So you keep thinking: How big does our intake funnel have to be to get two kilos of hydrogen per second?"

Storm considered. "I guess it depends on our velocity."

"You got it. And on the density. It's not high. I learned it in class. We're in the local fluff, the interstellar gas cloud. Actually, it's more like an antifluff, because the density here is a hundred times lower than in the rest of the galaxy. Bad luck—a single hydrogen atom per cubic centimeter. And it gets thinner toward Proxima. So you do the math in your head and you say to yourself: Something's wrong."

He took a big sip and slammed the glass on the table.

"Why?"

"Guess how big the funnel must be?"

"I think, two hundred kilometers. At least that's what the manual says. Seems huge."

"It's two thousand."

"What? Two thousand kilometer radius?"

Nygard nodded, smiling grimly.

"You must be mistaken. Multiplied wrong or something..."

"At first you get suspicious. *They made a mistake*, you think."

"Jason's conspiracy theory?"

The physicist laughed out loud. "I guess so."

"So what now?"

"If the calculation is correct, but the result is wrong—what do we conclude, commissioner?"

"Some assumption was wrong."

"Correct. Wrong is right!" Jason stuck his index finger into Storm's sternum. Storm giggled; the drinks were effective.

"I checked the archives. Local interstellar fluff, density, temperature, and so on. And it turns out that the numbers we learned in college are all wrong. The ridiculous Voyager probes were barely out of the heliosphere, I don't know what the age-old detectors fantasized about in terms of ion density—a hundred years later, thanks to Starshot, we knew better, it's not one atom of hydrogen per cubic centimeter, it's a hundred! And so the 200 km are enough for the vacuum cleaner. Fortunately, otherwise, EXODUS would never have been built. It wouldn't have worked. Starshot has opened our eyes."

They remained silent, concerned as they were.

"Not ours," Storm replied. "We were on ice."

The trend toward conspiracy theories kept the inspector occupied for the next few days. He couldn't put his finger on it, but he felt a connection to his former life. A feeling of unease, a suspicion that the attack on him was not only due to the cover-up of Carol's murder. Zigmund's connection to EUROFORCE, the wonderful fact that he, Storm, had been given a new life by this very EUROFORCE... Was there a bigger plan behind it?

He wanted to know more about CE-4 and met Jason again at their regular place.

"Slowly I am getting used to the whiskey here," the physicist mused.

"Yeah! It tastes a little bit more real than it did a year ago. They changed the recipe. Flavor enhancers, psychotropic drugs, maybe to calm us down."

"Typical conspiracy theory. Probably just your taste has changed. You know, nerves grow very slowly in the reconstructed body parts."

"Maybe that's the conspiracy theory. Maybe that's just what they make us believe. In reality, they're drugging us, the aliens."

Nygard laboriously scratched his head. "Occam's razor," he muttered.

"What?"

"William Ockham, thirteenth century, I think. Great philosopher. One should always choose the simplest explanation when there are several ones. Great advice."

"Let's have a toast to that!"

Which was the simpler explanation, Storm wondered. Was the more complicated one automatically a conspiracy theory?

After a while, the physicist said: "To a certain extent I understand the CE-4."

"You think they could be right?"

"No, I'm just saying that suspicion is normal. I had my doubts, too."

Storm was waiting.

"There are certain things..."

"The missing plasma light?"

"Yes, for example. But maybe that's okay. I can't check that. The funnel design is not available. Military secret. I was in the archives. It's a classified document and everything is blackened. Even if I had access, there's not much I could do with it. I'm not a plasma expert. It's damn complicated."

"Military secret sounds normal on a battleship. What else?"

"You remember the wind?"

"The one you missed?

"Yes, I got suspicious, but the anemometer proved me wrong. One can be mistaken. Like with the spectrometer." He shook his head in resignation, as if asking fate to forgive him for his helplessness.

"I often watch our sun," he went on. "Not on the monitor, but with the telescope. It's really nostalgic. It's orange-red from the Doppler effect. We're flying away from it, the spectrum is shifted to red. I wanted to measure the Fraunhofer lines."

"The what?"

"Not important. These are dark lines in the solar spectrum at a known wavelength. They're caused by absorption in the solar atmosphere. Their wavelength is changed by the Doppler effect. If you measure them you can calculate our speed."

"So what? Are we flying faster than the police allow?"

Nygard laughed and ordered a drink.

"There were no lines! I was going crazy. What's up, man? you think, then you take one more look at that flimsy spectrometer. Again, nothing—no lines! First, you doubt your perception. You repeat the measurement three, four times. It's true—no lines. Then you doubt your mind. Did you misunderstand something in your studies? You look up the Doppler factor, absorption lines, all good. Then you get suspicious. It can't be the sun. You think it's fake. *This* is CE-4 stage. When I was there, I could understand those guys."

"You think they're right?" insisted the inspector.

Nygard made a gesture of refusal and took his time. "Occam's razor," he finally proclaimed, looking at his counterpart as if expecting a student's answer at an exam.

Instead of answering, Storm concentrated on his glass.

"Come on, think! Either it's not our sun, or the spectrometer has a problem. Which is the simpler explanation?"

Now Storm understood. "The spectrometer was messed up?"

Nygard nodded approvingly. "The next day I took the spectrometer apart. The prism was filthy. Evidently, you can't see the narrow lines. Solution: soap, ultrasound, put it back in, and what do I see?"

"The—eh—Freudhofer lines?"

Nygard overlooked Storm's mispronunciation of Fraunhofer. "Bingo! A black line at about 580 nanometers. At first, I'm relieved. Great, that was just dirt on the prism. Then a shock: Damn, that's the wavelength of the sodium D line, we're not flying at all! Sodium D is at 589 nanometers and should be submerged in infrared, so much Doppler shifted at our presumed speed. So I look again because something appears strange. My mysterious line is in the green spectral range and not in the yellow as the sodium D. I repeat the measurement, and it's not at 589, it's at 562. It's the iron T line, which is exactly where it should be Doppler shifted when we're flying away from the sun at 55 percent of the speed of light."

"All real, no fakes?"

"All real. Physics is merciless."

"Now I'll tell you something. About fakes." Storm hesitated. "I hear voices sometimes."

"Cool. Mine, for one, I assume."

"No, you know … in my head."

"What do they tell you?"

"I can't understand them. Too quiet. Like through water. They come up from the floor."

"Well, that's a good sign. If you don't know what they say, you can't falsify it."

"Nonsense. I was going for Occam's knife."

Jason was thinking.

"I mean, we're always fantasizing all kinds of things," Storm continued. "Optical illusions. I'm sure there's acoustic illusions, too. Does that mean I'm crazy?"

"As long as the voices don't command you, you're just a harmless fool, I'd say." Only then he realized that Storm was serious with his question. "Your unconscious mind has cleverly arranged that you do not understand what they are saying," he conceded. "Well then: Is there someone underneath the ground, in other words in a vacuum? Is chief inspector Oliver Storm crazy? Or does his unconscious want to tell him something, but is prevented from doing so by his super-ego? May Occam's knife rule!"

When Storm replied nothing, Jason bent over, put his hand on his shoulder and said: "William of Ockham would, in all modesty, guess hypothesis three."

The Commissioner spent the next few days reading personnel files and tracking data. No one had been located near the arena during the time in question. It sounded frustrating, but for Storm, it was an investigative success, since it limited the search to people with time-outs. A comparison with those who had red paint resulted in 13 hits. He studied all 13 personnel files; 7 of them suggested a tendency to esotericism in the broadest sense. The suspect CE-4 was not among them, as he possessed no red color. He questioned all seven and found no evidence of perpetration. Nobody could be ruled out yet; even the suspected CE-4 could have borrowed the paint, perhaps from a conspiracy theory co-conspirator. Storm needed a motive. He realized that quickly. And that was hidden in the signature. What did Queen of Hearts mean?

He consulted the database of the EXODUS, which was designed similar to the Web. The first entry the system delivered was:

*Character from Alice in Wonderland. Children's book by Lewis Carroll, 1865.*
As he surfed on, he came across

*Alice in Wonderland Syndrome.*
*WHO International Classification H53.1 Subjective visual impairment.*
*Also known as* **Todd's syndrome** *or* **dysmetropsia***, is a neuropsychological condition that causes a distortion of* <u>perception</u>*. People may experience distortions in visual perception of objects such as appearing smaller (*<u>micropsia</u>*) or larger (*<u>macropsia</u>*), as well as altered acoustic perception, altered tactile perception and altered sense of time.*

Maybe that was it: they all had radiation-induced *Alice in Wonderland* syndrome. The blackouts, the feeling of looking down on oneself, on one's own distant body—micropsychia—often in half-sleep, but also in the waking phase. He knew that too. His sense of touch was altered. His skin still felt strange. He listened inside himself to see if his acoustic perception had changed, but could not detect anything.

He searched with interest. He found the Queen of Hearts in Chapter 8 of the children's book. He read it, found it more appropriate for crazy adults than for children. But what did he know about children in the nineteenth century? That had been another world, as strange as the one he had been thrown into by a tragic coincidence.

The Queen of Hearts was an evil empress who wanted to cut off everyone's head. Alice escaped this fate through courage. He read the passages again, and the queen now seemed to him less angry than disappointed and thirsty for revenge. Perhaps someone had done something terrible to her.

As to Zigmund... The Special Task Force of the past...

Following a hope, he selected from the personnel files those who had fallen victim to a bomb attack as civilians. Perhaps a Lorraine Bisset or a Marie Rückert or one of his serial murder victims were found, impossible as it sounded. The system found many cases, but as expected, there was no match for the six victims of his cold case so long ago. Again, he read too much into things.

He questioned Nygard about the details of the vandalism. He did not seem particularly distressed by the loss of his gyroscope. The problem, he said, was that now some two hundred grams of a special alloy had disappeared, which was a serious crime in a closed autonomous world that depended on recycling all matter.

"This is under martial law," he said dryly. "Ahlgrim, the asshole, has absolute power. The guy who did this will be court-martialed. No idea how this will turn out."

He looked at Storm carefully. "Maybe you're better off not finding who did this."

A few days before the shift ended, a message came from Earth. It was the answer to a question he'd asked three years ago.

*Archive inquiry Storm, Oliver. Personnel file extract from April 2077:*

*20.4.2077 assault. Severe injuries to extremities and abdomen from shrapnel bomb. Surgical reconstruction of the injured organs currently not possible.*

*24.4.2077 Admission to the EUROFORCE cryostasis project. Recommended by Prof. Dr. Peter Zigmund.*

*30.6.2078 Final report of the police investigation by Chief Commissioner Alice Falkenberg. Perpetrator and motive unknown. Case closed. Dissolution of the Special Task Force.*

*1.7.2186 Transfer of Oliver Storm in cryotank to earth orbit.*

*2.7.2205 Beginning of nanosurgical reconstruction.*

*1.9.2205 Admission to the waiting list for EXODUS II based on job description and psychogram. Recommender Bastien Calvin, CEO of MW Medical.*

*24.12.2210 Boarding EXODUS II.*

Zigmund, his former friend and rival, Zigmund the assassin had saved him. Without his intercession, he would never have been included in the secret project. And then there was Bastien Calvin, the boss of Zigmund's company, which had surprisingly survived for more than a century. Without them, he would have been dead a hundred and forty years. Storm's sarcasm stifled the amazement, the rising anger, despair, and gratitude. This called for a celebration. In his thoughts, he raised a glass to his two rescuers—a murderer and an unknown Bastien Calvin.

| | |
|---|---|
| Distance to destination | 1.463 ly |
| Time to arrival | 5 y 213 d |
| Distance from Earth | 2.780 ly |
| Speed | 0.490 c |
| Acceleration | −0.098 g |
| Slip correction | 240.00 d |

# Episode 11 Queen of Hearts

#news flash 1.7.2218
In the Riyadh assistance pact, China guarantees the sovereignty of Saudi
    Arabia. In return, Saudi Arabia cedes Jeddah to China. Chinese forces
    neutralize the Iranian militias.
A pole jump is imminent. The magnetic field has dropped to one-tenth of its
    historic value. With satellite communications and power grids collapsing
    across the northern hemisphere, external enemies are suspected everywhere.
    Russia is using the public mood for general mobilization.
© RG channel via moonbase Clarke II, sent 19.9.2215

Nordica became an assistant to David Müller, who seemed to have aged
years in the short time that Storm knew him. He was often silent and insecure;
sometimes he faltered in the lectures on the theory part, forgot to draw new
pictures on the screen, seemed absent and lacking in drive. Storm, who had
taken over the practical part of the schooling, approached him one day after a
lecture, when the others had already left and Nordica was putting away the
helmets.

"What do you want?" Müller fended him off. "I'm already 199 years old, if
you count officially. Some things are already difficult. The courses are
hard on me."

"We're all ancient," Storm said helplessly.

David laughed sarcastically. "Who knows," he mused. "It's all so unreal."
Softening, he put the helmet in his locker. "How lucky I am to have you. You
bring me back to reality, if there is such a thing. You make a good team."

He waved Nordica to him and laid hands on her shoulder and Storm's.

"A good team," he repeated happily. Nordica embraced David as a matter
of course.

A tender gesture that touched the inspector. How he would have liked to
be as spontaneous as this tall blond woman from the North. He had had
three or four liaisons in his life that had lasted a while before routine had
replaced the soft focus of love with the aberration-corrected lens of facts.
This woman was different. It was her reserved naturalness, her cautious
openness, her rational empathy, her disciplined gentleness. Or none of the
above, but once again a subjective Photoshop effect. Anyway, Nordica
pulled him up out of the drab emotionlessness that seemed to be a constant
part of his life. She brought him back to reality, as David Müller had said.

She was the first woman after more than a century who managed that, Storm realized, amused. And immediately afterward he was shocked by his sarcasm.

The interest was mutual. They started a relationship, which had a strangely distorted meaning on a European war fleet interstellar ship headed for Proxima Centauri. Between their duties on board and their daily routine, they had a wonderful time. She seduced him to creative experiments on the artl wall. He liked it, but his graffiti always reminded him of the flowcharts, time, and perpetrator diagrams that he had internalized as an investigator in such a way that they spoiled his artistic vein. Nordica joined his gymnastics group, and the two of them enjoyed themselves in the Belt with fantastic flying figures in weightlessness. The Jumper became a place of sensual ecstasy.

After the dreamtime with Carol, Oliver Storm had always found sex to be fraught with fear, which he had learned to cover up with sarcasm. Foreplay, cleansing procedures, the embarrassment of undressing, the presentation of bodies so far removed from internalized ideals, disillusionment in daylight— all this prevented a seamless surrender and letting oneself go, as it should be in a love relationship, he used to think.

Here it was different: the uniforms almost fell off their new bodies by themselves, which were clean and immaculate. Apparently, some of the genes had been reprogrammed during the reconstruction. There was no perspiration here, no hygienic preparations were necessary, the movements were playfully safe and of floating elegance in the low gravity.

Soon, the sparse time-outs and escapades in the jumper were no longer enough for their intimate get-togethers. Nordica suggested trying VR.2 to get around the time limit. As an expert in terahertz technology, it was easy for her to construct an environment from archive material that was at home on an Earth long before climate change had struck. A South Sea paradise under palm trees before the demise of the islands, a châlet in the Alps before the melting of the glaciers, the Burj Khalifa before it was destroyed by Christian fundamentalists, the Louvre before it was plundered, and much more. They often retreated to their private virtual reality—VR.2b he called it—after work, enjoying enchanted landscapes, cultural monuments, and luxury. It was so lifelike that sometimes, when Storm took off his helmet and returned to the sober reality of the EXODUS, he found *this* to be virtual. Here a veil hung over things, they seemed to have a weak second existence, as if a parallel universe was reaching out to them.

"You get used to VR.2," he said. "If we didn't know it was fake, we might think it was real. At least sometimes."

"What do you mean, *real?*" she replied.

"The physical reality. The one Jason keeps telling us about."

"Oh, yeah? Atoms, maybe? Can you touch them? Can you see them?"

"Sure I can. There are electron microscopes."

"You can do that in VR.2 if you program the camera. Very easy task."

Storm was silent. Philosophy wasn't his favorite subject.

"I'm afraid your definition is a dead end," she persisted. "The *thing-in-itself* is inaccessible to us. We have only our perceptions. We perceive it as being true, we just assume that it is true. All illusionists work with perception to fool the audience. Speaking of illusions: the many optical illusions we had fun with as children—parallel lines that seem to converge, figures of different sizes that are the same size when you measure them,... I never understood why one talks about optical *illusions*. It's what our neural computer processes from inherently meaningless optical input. It's not a delusion, it's true. Perceived."

She was so smart. He was sure that she could help him in his cold case. He was often close to initiating her into his findings—the serial murders, the false lead to Carol Beauclere, his lover from a lost time. And finally Zigmund, the mastermind behind the bombings, who had hurled them both into this new world. Should he be grateful for survival or should he be angry at being manipulated? Breathless, he felt the absurdity of the World Theater. But he never got around to confiding in Nordica; maybe he didn't want to jeopardize their love with disgusting details like cut off heads, maybe he didn't want to be questioned about Carol, didn't want to mention his old friendship with Zigmund, or it was simply cowardice.

Storm continued his investigation in the gyroscope case. He squeezed all the people who had had time-out on the night in question, read personnel files, interviewed, checked alibis by interviewing their friends and acquaintances, cross-checked the claims with VidCam footage from the archives. In the end, there were five people left who had no alibi. Klaus Amann, the already suspect air-conditioning technician, his friend Sean O'Brian (a younger man with whom Amann had an intimate relationship), a technician from engine control, and, curiously, Jason Nygard and David Müller. Amann and O'Brian claimed to have had intimate intercourse during the time in question, but the other three had been alone. These three had no discernible motive (the technician didn't, Nygard and Müller certainly didn't).

The inspector, therefore, concentrated on O'Brian. The Irishman had fallen victim to his own bomb at the age of sixteen while preparing an attack on the British occupying forces and had been accepted into the EUROFORCE regeneration program. He painted, read, and played music in his spare time, and except for Amann and two other CE-4s he had hardly any contacts. An aesthete, highly sensitive, and introverted. Storm asked him straight out if

"Queen of Hearts" meant anything to him. Sean came up with a retro song from the twentieth century. He thought about it for a second and hummed a melody that was unfamiliar to Storm.

Did he know Alice in Wonderland, the nineteenth-century book by Lewis Carroll? No, he didn't, he responded. By the way, he couldn't imagine why someone would want to destroy this physicist's magic device that nobody understood anyway.

*But maybe that was it,* the inspector thought. Maybe it was the device's incomprehensibility that made it so dangerous. He stopped the questioning.

Somehow this guy seemed too innocent. Storm checked the downloads from the library. Under Lewis Carroll, Alice in Wonderland, he found only two recent downloads, made by himself and Sean O'Brian. The case was clear. He ordered the two intimates and confronted them with the facts. Amann was stunned, O'Brian on the verge of tears. He admitted that he'd read the book, but only because of his interest in nineteenth-century literature. He quoted Storm a list of authors from the period: Dickens, the Bronte sisters, Kipling, Stevenson, Thackeray, Wells, etc. He was ready to retell any novel (or almost any novel) from that era on demand. But why had he denied knowing Lewis Carroll's book before? Fear of being suspected was the plausible answer.

The final report presented the inspector with a problem. There was no confession, only circumstantial evidence, and no proof. Had O'Brian planned the crime with his partner, or was Amann actually innocent? Amann's bewilderment seemed real. So maybe they were telling the truth. Rarely had Storm been so insecure. And the reference to the Queen of Hearts made no sense. He discussed the matter with Nordica.

"The Queen of Hearts is cruel and evil," she mused. "The CE-4s are in a hopeless situation. They've been kidnapped, vivisected, tortured, deceived, pilloried. And whose fault is that?"

"The aliens'."

"And the Captain, who ignores their concerns. He may be the Queen of Hearts."

"This pillory keeps me busy. Can the world change like this in a century?"

"What do you think?"

"A return to the Middle Ages..."

She pursed her lips. "No, it's rather a recent evolution. We had similar punishments in our time. For the slightest deviation from mainstream opinion, people were denounced in the media. It may not have cost them their lives, but it cost them their careers. They were pariahs, outlaws. It just seemed more fancy. They called it shaming a century ago."

The inspector was deep in his thoughts. Was Ahlgrim merely a natural progression from the self-righteous tyrants of the twenty-first century? A perfected Zigmund...

Then he heard a muffled murmur, but it wasn't spoken by Nordica. It came from the floor below: "O'Brian. He's bought it."

Storm tilted his head and listened to the phantom voice.

"The clue..."

"Wait!" Storm raised his hand to stop Nordica's speech.

"Stress..." Another whisper came out of the ground, hard to understand. Hurried steps became distant.

"What was that?"

"What was what?" Nordica looked at him without understanding.

"Someone said 'O'Brian has bought it."

She looked at him skeptically.

"It was very quiet," he justified himself, pointing down. "And the steps..."

"There's a vacuum down there. Deep space." There was an embarrassed silence. "You're overworked. The case is haunting you, deceiving you. It's your guilty conscience about O'Brian."

Storm rubbed his eyes as if he wanted to wake up. "You're right. I should look at this more professionally. It's a job, nothing else."

"Sometimes I think I hear voices too," she acknowledged sympathetically, shaking her head. "This damn spaceship is going to destroy us."

*How right you are*, he mused.

"The Queen of Hearts clue is a cry for help," she picked up the thread of their earlier conversation. "It's an allegory for the fate of the CE-4s. Alice is likewise in an illusory world. She fights with logic and courage against all the strange sorceries in an alien world. It is a rebellion against fake facts. You destroy the gyroscope, a symbol of deception. This—what O'Brian called it—this physicist's magic device that nobody understands."

A vague idea flashed through Storm's mind.

"David said something strange that I can't get out of my head. Everything seemed unreal to him. As if everything was an illusion."

She nodded. "Weird. And he's changed. He's so... lifeless now. As if the fate of the crew no longer concerns him."

"But we're bringing him back to reality, he told us. Isn't that great?"

The next day the Commissioner finished his report. It included Nordica's evaluation, all interviews, research, analysis of conversations. He ended by stating that there were certain indications of the culpability of Amann or O'Brian or both, but that in the absence of a confession or evidence, the only possible solution was an acquittal *in dubio pro reo*.

He summoned Amann and O'Brian and informed them of the results of his investigation. He would recommend an acquittal.

Amann seemed relieved; his hostility turned into reproachful relief during the conversation. O'Brian didn't seem to care about any of this anymore. He seemed reticent, and his gaze was dull. When he left Storm's office, his footsteps echoed like a zombie's.

*It's true, O'Brian had bought it*, thought Storm.

Days later Thelen ordered him onto the bridge. They sat in the small meeting room where Silke had been shot. This had happened in a VR environment, but nevertheless, Storm had an unpleasant déjà-vu. He felt uncomfortable, as if agitators could rush in at any time and detonate a bomb.

Thelen swayed his head in doubt.

"Your *in dubio pro reo* proves that you have studied jurisprudence with ardor keen," he mocked. "But I do not need to remind you that we are at war. Roman law is irrelevant here. The sentence depends on how plausible your evidence is."

The Commissioner pursed his lips and considered.

"The last exercise was very realistic," Thelen replied. "Especially the verdict."

Storm understood. Mutiny, sabotage, conspiracy … The verdict would be harsh. He thought long and hard to reconcile his feelings with reason.

"The circumstantial evidence is not very plausible," he claimed, although without much conviction.

"I'll pass this on to the Captain. If he needs more specifics, be prepared."

But Ahlgrim didn't want more specifics. A few days later, he ordered the crew into the arena. This time, it wasn't a Coriolis-style leap. He came strutting along the Promenade. Behind him, Amann and O'Brian were paraded in handcuffs. Ahlgrim stood in front of the podium with the prisoners.

"You know what this is about. The charges are sabotage and conspiracy. The destruction of equipment is a capital crime."

Armed soldiers watched the crowd, which murmured uneasily.

"I have received the final report. There are five suspects. Some evidence points to Amann and O'Brian, but the commissioner is not sure. The accused deny the crime. There is no evidence. Therefore, only pillory for them. The others go free."

Storm was relieved and ashamed. Nygard had proved right. It would have been better not to find any suspects. He sneaked along the small promenade far away, past the Column of Shame. To his relief, nothing degrading happened. No one came along to humiliate the two wretched creatures.

The second day, Nordica brought them food. Apart from her, none of the crew reacted. The arena remained empty, the area was avoided. *Cowards,*

Storm thought. It was better to always assume the bad in people. Then he remembered that he hadn't done anything good himself. He left in shame. His cynicism let him down.

| | |
|---|---|
| Distance to destination | 1.008 ly |
| Time to arrival | 4 y 213 d |
| Distance from Earth | 3.235 ly |
| Speed | 0.419 c |
| Acceleration | −0.098 g |
| Slip correction | 280.00 d |

## Episode 12 Shangri-La

#news flash 1.7.2219

In the Treaty of Pjönyang, North Korea and the USA agree on mutual military support. The USA supplies fusion technology to North Korea. The Seoul Supplementary Treaty ends the inter-Korean conflict.

The imminent geomagnetic reversal triggers worldwide volcanic eruptions. Ash clouds of Eyjafjallajökull, Nyiragongo, Vesuvius, and Krakatau spread in the stratosphere. A nuclear winter is looming.

Russian troops advance as far as Berlin and Prague.

© RG channel via moonbase Clarke II, sent 6.4.2216

The plasma flicker in front of the bow window went out, the drag pressure decreased, and the landing craft tilted its nose into gliding flight

"Mayflower, you are well on your way. Flight parameters are in range. I'm taking you into the entry sequence now."

"Copy that, Mission Control," Storm confirmed. The shuttle rolled into a slight left turn. Where the night-dark sky touched the horizon, a green-gold band glowed, precursor of sunrise. As they headed for the terminator, Proxima Centauri, the central star of their new home, came into view deep in the eastern horizon. A pink sky stretched across a rocky landscape whose skyline was barely curved from 20 kilometers above.

The descent led them along the shadow border toward the south. The jagged rocks dissolved into mighty domes and canyons.

"I can see Cape Cod!" Dana reported from the copilot's seat. Directly in front of them on the border between day and night, a wide plain opened up. The long shadows of a mountain ridge to the west pointed to a natural landing strip, a smooth plateau. In the low morning sun, every elevation of the ground was clearly visible.

Storm confirmed the sighting of the landing position and asked for permission to land manually.

For the last five hundred meters of altitude, the Mayflower was shaken by strong crosswinds. The boat rolled and yawed, and the seemingly slippery rock, on which they sank down far too quickly when a downwind caught them, was covered with numerous stones, from fist to head size. The landing was anything but smooth.

"Mission Control, what's our status?" Storm asked.

"All right. You'll have enough juice to get back. Slight problem. Your landing gear is damaged."

"Details?"

"The nose wheel is showing a defect."

"We'll check it out. The area looks calm, no sandstorms, no ripples. Westerly winds, 40 kph. Can we get out?"

"Affirmative. Have fun in the New World."

They got off without helmets. The strong westerly winds blew cool air into their faces. As Storm examined the nose wheel, Dana looked out into the strange landscape. The huge sun hung low in the west in a flaming sky, which changed to the zenith to dusky pink and further to the east into a soft green. Veil clouds loomed over the distant mountain ridge that cast its shadow from the sunset onto the plain. A tiny moon with low albedo was rising into the green sky.

Storm straightened up. "The wheel is still on, but the tire is gone."

"Amazing," Dana whispered. She hadn't heard him. "Look at this. It didn't look like this on the videos from the satellite. It's otherworldly."

"The satellite doesn't understand beauty." Storm tried to adjust to her mood.

The satellite, which had been sending data for over ten years, showed a planet whose year only lasted twelve and a half days. The rotation was bound: Atlantis always showed its sun the same side. Slightly smaller than Earth, Atlantis had a 20% lower gravity, a magnetic field about twice as strong as on Earth, and a breathable atmosphere. There was water that was deposited on the night side at temperatures down to minus 100 degrees Celsius in mighty glaciers and seas of ice. The day side was hot, dry, and rocky. Only close to the terminator, where the sun was low on the horizon, the climate was bearable with temperatures between –10 and +25 degrees Celsius. There were large lakes, river landscapes, and strange plants. The warm air rising on the day side created upwinds at the shaded border and at the same time ensured a constant air–water cycle.

Close to the high plateau, which they named Cape Cod after the landing site of the historic Mayflower, wide valleys and gorges opened up, and waterfalls plunged into them. There was vegetation—mosses and lichens on the shady western slopes, small plants, shrubs and gnarled trees, always nestled against rocks. The rotation axis of the planet was inclined by 15 degrees against the ecliptic. The planet's orbit was strongly elliptic with an orbital eccentricity of 0.11. During one Atlantis year, Proxima moved on an elliptical orbit across the western sky. It stood six days—half of the "year"—below the horizon. This was enough to create a weather cycle that reminded Storm of hasty seasons. The flora had adapted over millions of years and produced short-lived flowers. How the fast-growing plants survived the occasional flares of the central star with violent particle storms was puzzling. Satellite data provided no information about this. And there was no sign of animal life.

"How bad is it?" Dana asked.

"The nose tire is shot, but we can start on the wheel if we engage VTO. We've got plenty of juice."

Storm sent a situation report to EXODUS and after a short consultation, David gave him the OK for the planned reconnaissance trip. Storm let the E-Buggy roll out of the charging bay, got into the saddle, and checked the battery, camera, and radio.

"Okay, the plan stands. Meet here in an hour, as planned," he said. "Protocol check, please."

"The Mayflower remains on standby, scheduled takeoff in one hour," Dana reported. "If you're not there by then and there's no radio contact, I'm taking off anyway. Flight over base camp, attempt to locate you, return to EXODUS."

Storm gave her a farewell salute and quickly rolled south, where the terrain gently sloped and scattered bushes lined the runway like radio beacons.

The buggy rode stable on the track and easily handled the bumps and the strong gusts. It went steadily downhill. After a short time, the lovely valley they'd chosen as the base of their colony—and named Shangri-La—came into view. Grass and bushes grew there, nourished by several streams that crossed the valley from northwest to southeast. The sun blazed in the west over a cliff that towered several hundred meters high. From the snow-covered ridge, a silvery band wound its way down into the depths—a waterfall, the source of the streams. The Grampians, as they had called the mountain range, protected the valley from storms and tempests that had smoothed the plateau of Cape Cod over the millennia. To the east, a gentle range of hills rose, warmed by the low sun.

At the bottom of the valley, there was almost no wind, and the air was pleasantly warm. The cargo module from EXODUS, which had landed a few days earlier, had unloaded automatically. On the bank of a stream stood the accommodation container, and next to it were the weather station, the directional antenna, the superconducting coils of the particle protection, and two buggies. Storm dismounted, looked around, circled the station. Everything was quiet.

"I'm at position B," he told Dana.

"Copy that. Status report?"

"Everything's going according to plan, the container is in place. I'm checking the station."

He jiggled the antenna and checked the moving parts of the weather station. The anemometer was spinning sluggishly, the panel showed easterly winds at 5 km/h, temperature 21 degrees Celsius, humidity 60%. Good visibility.

Storm looked around. Above the Grampians, cumulus clouds hung in constant motion. The wind blew them over the ridge, where they formed as they rose and soon dried up again when they met the warmer layers of air on the day side. A soft, distant howling of the storm was the only sound in this strangely alien landscape that stretched far to the south and north. There was nothing here but grass and bushes.

The inspector unlocked the container door and entered. He found the containers for the soil, water, and plant samples and the tools.

"Taking the samples now."

"Copy that," Dana responded. "Everything is quiet here."

He took the water sample from the stream that ran alongside the container. He strolled along the watercourse in search of special plants, but he found only the spoon-shaped grass, mostly sheltered by larger rocks. He tore out some stalks, tucked them away, moved away from the water, up a hill. At the highest point, he took a soil sample with a stabbing tube. Back at the buggy, he tucked away his boxes. He brought the tools back into the container. When he closed the toolbox, something fell to the ground. It glowed faintly - a mablet that said:

*Permission to override VR.2 simulation Atlantis.*
*To Commissioner Oliver Storm.*

*Please read this text carefully. The device will self-destruct in three minutes.*

*Europol needs your help in a cold case.*

- *Please tell us if you had any observations prior to the attack on you that might help us solve the series of murders you investigated as head of the special task force.*
- *Question Nordica Henderson about illegal connections between THZ S.A.R.L. and EUROFORCE or other government agencies.*

*Please proceed with discretion. Do not talk to anyone about this.*
*There will be a major GCR event on July 16 at 0:10. Please be alone in your cabin at that time. We will contact you.*

The inspector read the text several times before the tray in his hand withered like a rose in autumn; dust trickled to the ground, nothing reminded of the message.

What was that? A joke by David Müller? But it hadn't sounded like a joke. Confused, Storm locked the container behind him, memorizing the contact data. Then he checked the time and radioed Dana: "All samples stowed away. I'm on my way back."

"Copy that. It's—" The rest of her answer died in a swelling rumble. A cool wind blew from the top of the hill, at first restrained, then more violent, and that's when he spotted it: an almost vertical swell, a tsunami rolling down the ravine. He jumped on the buggy and tried to avoid the wall of water crashing down the hillside, but it was too late. The wave washed over him, tore him off his saddle, and dragged him away. Behind him, he saw the container bouncing on the waves. Over bumps, boulders, and rapids he went, whirling, upside-down, tumbling, until the current slowed to a leisurely flow and the journey ended in a lake.

The water was cold, it smelled like snow. He swam to the nearby shore, climbed up a mossy slope and looked around. The water continued roaring into the lake, bearing bushes and plants. The container, half-submerged, sloshed languidly along at some distance. Looking upstream, he recognized the cause of the tsunami: Dark clouds hung above the western cliff, from which strips of heavy rain fell. The graceful waterfall that fed the creek had turned into a tapestry of thundering floods.

Storm was completely drenched, but that didn't bother him much. His skin sensitivity was limited in VR.2—an advantage in his current situation. The buggy with the samples was on the bottom of the lake. Its microphone and earplug had been washed away. So he was unable to inform Dana, and she would return to EXODUS soon as agreed, without him.

There was nothing he could do. The exit point was in the settlement near the antenna. In vain and against all experience he tried to take off the Rift, then to take off his helmet, but of course, his virtual hands only felt his virtual body—head, ears, eyes, his hair—everything as it should be with good software. So he climbed up the hill, lay down in the shade of a lonely bush, used a stone as a headrest, and looked at the landscape. Upriver, the creek had found its way back to its bed. On the lake, the container was about to sink. Upstream, Niagara Falls subsided, the clouds above the ridge dissipated. In the southeast, on the lakeshore, the landscape was indistinctly gray. He thought he saw a cross rising out of the water in the distance—as if covered in fog. Further to the east, the view cleared up and revealed hills and rocks. Even to the south the air was clear, the sunlit cliff stretched to the horizon, only the end of the small lake was covered in fog. Curious, he set off along the shore, the waves sloshing over the rocky embankment.

Halfway there, he heard the engine of the Mayflower. In a wide arc, it moved across the lake and disappeared in the cloud. The roar of the engine suddenly stopped. It was dead quiet—as if a wall had swallowed the shuttle. The cross, he recognized as he approached, did not protrude from the water, it hung in the air, small, formed of thin black wire. As he approached, details appeared: two more crosses left and right, a line. Soon he stood right in front of it; just above eye level, the cross hung in the gray. A dashed horizontal line connected it to its neighbors and others. All of them hung in the air at regular intervals as if on a pasture fence.

Storm reached out for the wire, carefully groping for it as if he feared an electric shock. He touched a cold gray wall. The wire was a line drawn on the wall with cross-shaped markings that extended to both sides. Next to it were numbers and symbols. He went closer. Atlantis N48.208411 E16.373471 SO was written next to the large marker.

He had reached the edge of the world.

***

David's analysis was devastating. They both should have seen the threatening cloud over the ridge when the Mayflower landed. The waterfall had already shown irregularities when Storm took off. His carelessness, when the tsunami had announced itself by wind and rumbling, was due to the message on the mablet, but that was not something Storm could reveal, and it was no excuse. He had simply messed it up.

Discouraged, he withdrew to his apartment to, as he said, go through everything carefully for the next exercise.

Who had sent him the message on the simulated mablet? It was about his case, about the task force and about Zigmund, after more than a century a veritable cold case. Was there a Europol v-person on board? He filtered the files of the crew members according to every conceivable criterion. Age, place of residence, profession, history, but he found nothing. But he found something else: There were no civilians on board who had been included in the EUROFORCE reconstruction program before him. All of them had fallen victim to an assassination attempt or had been injured at the same time or later. And all of them had come out of the tanks after the launch of EXODUS II. Only the officers were active from the beginning.

Maybe the earlier cases had been on EXODUS I. He requested its crew list from the database. The system responded with

*Access denied. Classified file.*

Obviously, someone wanted to hide the information.

It was a dismal failure for an investigator used to success. Storm brooded for days looking for another access. Finally, he remembered the Forensic Time Machine, the software that had virtually taken him back in time in another century. Luckily, the program was in the starship's system. He waited for his time-out; in the training room, he put on his helmet and VR goggles, activated the FTM program and entered EXODUS II, Belt, July 1, 2219, 17:00.

It buzzed, then the big promenade built up before his eyes. Two people jogged through it. The flat screen at the front of the houses showed the dates of the journey, and below them the news of July 1:

#news flash 1.7.2219
In the Treaty of Pjönyang, North Korea and the USA agree on mutual military support. The USA supplies fusion technology to North Korea. The Seoul Supplementary Treaty ends the inter-Korean conflict.
The imminent Pole Jump triggers worldwide volcanic eruptions. Ash clouds of Eyjafjallajökull, Nyiragongo, Vesuvius, and Krakatau spread in the stratosphere. A nuclear winter is looming.
Russian troops advance as far as Berlin and Prague.
© RG channel via moonbase Clarke II, sent 6.4.2216

He had seen that on the screen. FTM worked.

He went back to 1.1.2210, the launch date. The entire crew was gathered in the arena. Storm zoomed in. It was Crew 1, he didn't know any of them. Padmé sat in the front row. Molander spoke. Storm explored the Belt, entered the medical unit in the igloo unhindered. There—the cell he'd woken up in, empty. He went through the wall into the tank area. There the sleepers lay in reconstruction. The picture was grainy like in old newspaper photos— Nordica, Zimmermann, O'Brian he recognized, the other bodies were only shadows. Storm's mangled body was here too, somewhere in this room, plagued by nightmares while he waited for his resurrection. He entered the cryostasis area and the image turned gray. No information available on the passengers, that meant.

Back to the Belt. Molander was still talking. Storm glided, steering with hand movements through the habitat to the bridge. Horst Thelen sat at the control desk. The screen, which resembled a panoramic window, showed a huge moon drifting sluggishly through the field of view and disappearing sideways, as if in no hurry to release the EXODUS from the libration point into space. A starry sky with residual light replaced it, the band of the Milky Way shone above ground. And then the Earth came into view, dazzling and powerful, a sphere shimmering in the blue of false hope. Storm felt a choking in his throat. *We betrayed you,* he thought. Before homesickness overtook him, he zoomed away.

Then he asked for the coordinates of EXODUS I as a new target, without much hope of success. But the FTM reacted: He drifted through the Belt, the habitat, the bridge and left the ship through the bow. Alpha Centauri shone before him; he almost felt the pull of the vacuum as the software catapulted him to the required point in space in a ghostly animation. The sister ship, a tiny blue dot, became blindingly bright. As the red-hot lotus leaves filled the field of vision of the goggles, he flew unhindered through the fusion fire of the engine, on to habitat. The pipeline punctured him on his way inside.

And then the image turned gray. He tried to orientate himself in the void, zoomed in different directions, but it didn't help. *No information,* it flashed in the intrinsic grey. Only when he steered himself back into the vacuum did the EXODUS I reappear in the VR glasses.

Disappointed, he ended the FTM and took off his helmet and Rift. No information. Maybe they didn't have any, or they didn't want to give it to him. To him or to others who had searched before...

He opened the log file from the Forensic Time Machine. It was a short list. At the top was the last activity: his name, location, start and end times, followed by the coordinates he searched for. He scrolled down. Names of crew members, strangely enough mostly from crew 7, Jason Nygard was there, David Müller, then came unknown names, maybe twenty entries in total. He went back through the names, trying to associate something with each one. His eyes got stuck on Eckermann, Sandra.

Somehow, he must have known the name. Not from crew 7, he was sure of that. He looked closer—EXODUS I was written next to her name. No doubt: she was a crew member of I, not II. Then he remembered: The special task force. Sandra Eckermann, Master of Biology, the third victim in the series of murders. Almost victim, if there was such a term; because after the attack, she became a crew member on the EXODUS I and had used the Forensic Time

Machine in search of the truth, before she died for the second time in the explosion of the spaceship.

| | |
|---|---|
| Distance to destination | 0.996 ly |
| Time to arrival | 4 y 202 d |
| Distance from Earth | 3.248 ly |
| Speed | 0.417 c |
| Acceleration | −0.098 g |
| Slip correction | 281.03 d |

## Episode 13 Attack

#news flash 12.7.2219
Chinese armed forces crush uprisings in the European autonomous trade zones. Over a hundred thousand dead in Spain and Italy. Europe moves troops to its southern fortifications along the Loire, Alps, and Danube.
The European Court of Justice orders the expropriation of Chinese property in Europe following the incidents in the autonomous regions.
© RG channel via moonbase Clarke II, sent 12.4.2216

The new shift began with sad news. O'Brian had not awakened from cryostasis. The oracle's whispered prediction had come true. The sensitive Irishman had been devastated after the interrogation, and after that, he had been apathetic, lost in despair. *Perhaps,* Storm mused, *my unconscious was trying to point me out, to warn me when I heard the voices.* They now had three casualties in the crew—two death sentences before his time, and the Irishman. Too much, but not much for the ten years since launch.

During the night Storm jogged on the small promenade to clear his head. It was dark, the pipeline glowed dimly. The longer he thought about it, the stranger the events seemed to him. The EXODUS I explosion, secret lists, conspiracy theories. Even familiar things like GCR events, flashes of light, the stale taste of certain foods, the strange skin feeling... suddenly everything seemed suspicious to him. Something was going on here. Were the CE-4 right?

In moments like these, he felt at the mercy of fate, like a proton sucked into the fusion chamber by the magnetic field of the EXODUS. As he drifted toward an inevitable endpoint, he waited for the player in whose computer game he felt he was trapped, to push the reset button that would wake him up in his previous life.

The reopened cold case; Zigmund's perverted murders of the RAVEs in revenge for the bomb, a mirror punishment. The attack on Storm for finding Carol's body.

What did Nordica have to do with it? She was involved in the development of terahertz technology. What did she know that couldn't be disclosed? What was the connection to Zigmund? The coincidences, oddities, and correlations were strange. Coincidences are the wink of God's eye, a poet had once said.

With long jumps he sailed over the promenade, up to half the height of the igloo, slowly sinking to the ground, while the EXODUS turned under him. On the large promenade there were few joggers; the small one he had for himself. In the igloo an open door, weak pulsating light behind it.

That was unusual—the tank area was supposed to be closed. Carefully he peered inside. A flashing red light cast shadows across the room. It came from the right. Out of the corner of his eye, he noticed a movement there, a scurrying like the wings of a bat. He carefully pushed off, hovering in the low centrifugal force between the tanks up to the wall where *EXIT REQUEST* flashed red every second. The massive security door to the cryostasis area was left open a little bit. *Hibernation area—no entry.*

Storm looked around, hesitated briefly, then pulled the door open. Someone grabbed him from behind, threw a sling over his head, and pulled it tight around his neck.

He knew that if he didn't counter the attacker, he would lose consciousness within ten or twenty seconds and die within two minutes. He felt the pressure of the noose, but he could still breathe. The training was to his advantage—he curled up, grabbed the attacker headfirst by his upper arms, and threw him forward. In normal gravity, this would not work, but here it was easy to move two inert masses of seventy kilos that had no support, against each other. The attacker was not prepared for this defense, came forward rotating above Storm's head, and the sling came loose.

Storm was now behind the attacker. He wrapped his left arm around the man's neck and pushed off. Together they headed for the security door. An instant later the enemy's head crashed into the metal frame, and his body went limp. Storm let the man slide to the ground. Then he freed himself from the wire sling, which floated to the ground in the low centrifugal force, and he bent over the motionless body. Athletic, medium height, uniform. A mask of fine gauze covered his face.

"Who the hell are you?" Storm muttered and grabbed the mask. Then he felt a kick in the stomach and sailed away helplessly, while the revived person escaped through the door on all fours, pulling it behind him and locking it. When Storm was finally able to brake his out-of-control drifting, the flashing light had gone out. He tried in vain to open the door, looking for the murder weapon, for traces of blood. Next to the bulkhead, he found a piece of gauze. He had torn it off the attacker's face. He examined the fragment, which turned out to be a flexible plastic with embossed gauze strips. The piece of cloth had a number 137 written on it. Next to it, on the edge of the tear, he could decipher the beginning of a word: "Ca..."

This was the second time that someone had tried to get rid of him. He wondered for a long time whether he should report the incident—an unknown attacker from the sleeper area had gained access to the tanks. His report would sound like that of a CE-4. There were no traces of the attack except for that piece of plastic. Storm put it in his pocket. For the time being, he would keep the case secret.

The next day, he went through the passenger list. Number 137 belonged to a Bastien Calvin, in cryostasis since 2210. That same Calvin, CEO of MW Medical, who had put him, Oliver Storm, on the waiting list for the EXODUS. The few personal details told an interesting story: studies in philosophy, eugenics, and robotics, professor at the ETH Zurich, military training, then laboratory manager of MW Medical Inc. He was a distant successor to Zigmund as director of his company, which still existed after 137 years. Strange coincidences: 137 years, the number 137... the initials Calvin, Bastien—Carol Beauclaire...

He pondered for days over the attack and the strange coincidences, discarding one scenario after another and decided not to involve others in his deliberations. Collaboration wouldn't work; they would think he was crazy or worse, if they were among the conspirators, he would give himself away.

After that, he kept everything to himself. He told Nordica only about his work in the homicide squad, told her about the serial murders, about the search for the perpetrator, about his suspicions, the wrong trail, the assassination attempt on him, just when he had discovered the ceramic stove in which a corpse was probably walled in. He said "probably" even though he was sure.

"I think you were a victim of the same criminal," he concluded.

Nordica didn't dispute the point. In fact, she readily agreed with Storm. "He wanted to get rid of both of us because we knew something that could have brought him down." She had obviously already come to similar conclusions. It was good to know that he wasn't the only one with an enigmatic past.

"We can't do anything about it anymore, though," she continued. "He's long since buried. Your case is closed, one way or another."

"Yes and no. We still have contact with Earth—the directional radio is working again." He pointed his thumbs up. Storm wanted to show his eagerness for pursuing the investigation, but at the same time, he didn't want to reveal the vanishing message he'd seen in the Atlantis simulation, the directive from Europol urging him to question Nordica.

"We could solve a cold case. Do you remember anything unusual that happened just before the assassination attempt on you? An observation, a phone call, something at the lab. An irregularity in the experimental setups, perhaps? A little something you didn't give any weight to?"

She tilted her head back and closed her eyes. That was her way of thinking intensely. He liked that.

"There was something about a week before the attack. There was this guy, Zigmund, and this guy from the Ministry."

"Say that again. What was his name?"

"I only knew the Ministry type by sight. Bald guy with glasses, high voice. He was often there when they discussed the project."

"No, the other one."

"Zigmund? He was a consultant, hired by the Minister to advise us. He was the contact person."

"Zigmund? Prof. Dr. Peter Zigmund? Of MW Medical?"

"Yes."

"This can't be true," he muttered, as if he didn't want to believe what he already knew.

"Why...?"

"He was our prime suspect."

"You mean, he did the bombings...?"

"I don't know if he himself... Anyway, he was definitely involved."

Storm poured himself a Scotch, even though it was early in the morning. His hands were shaking. He took small sips to calm himself down. After a while, he was able to talk. "Tell me. What happened?"

"It was about the big project. Transcranial stimulation of the afferent nerves with terahertz waves. Top secret. Actually, what we're using here with David— we had version VR.1. Our contractor was the Ministry of Peace, but we knew the secret service was behind it. Of course, they never appeared, everything was handled by this Zigmund. I was in the next room, it was late at night. I'd been evaluating some tests. The door was open, they hadn't noticed me.

"Zigmund wanted our software to be more realistic. The waves were too imprecisely focused on the homunculus in the cerebral cortex. And then it was about new test subjects for VR.1. It sounded like they were running out of them. The Ministry man asked Zigmund for supplies. He wanted *good brains*, yes, that's what he called it. It was a strange phrase, that's why I remember it so well. Good brains, he said, and from civilians. He'd had enough of the military. Zigmund promised to deliver. They also discussed cost. Zigmund wanted more money for the new ones he was to deliver, he demanded a horrendous bounty."

She interrupted herself, took a sip.

"Bounty—how that sounded... I think it was twenty thousand. A badass headhunter. It got pretty loud, because the one from the ministry thought it was too much. Zigmund said something about high medical expenses and the risk. And then a name came up, a contact at EUROFORCE who was supposed to do the logistics."

"And then what?"

"That's all I heard. I didn't want to keep listening and be suspected of eavesdropping. I finished my work, said goodbye and left. The next day, the boss called me in. He was very nervous, babbling something about classified projects, sensitive data, secret service, and so on. He said under no circumstances was I allowed to talk to anyone about what I'd heard, not even hints. If anything came up about the meeting, he was finished. But I knew that anyway, so no problem, I said."

"Do you remember the name? The contact at EUROFORCE."

"Yes, funny name, Hasselborn was his name."

"Hasselborn?" That was the drone pilot who was supposedly monitoring the Doomsday adepts—in another life.

They now shared a secret and felt better equipped to deal with the increasing outbreaks of suspicion, rumor, and recrimination—consequences of Ahlgrim's barbaric practices—the consequences of the state of war. *Perhaps shared secrets are the key to solidarity*, thought Storm in moments of uncertainty, when reality threatened to dissolve. The time with Nordica was the only thing that made him believe he was real.

On the evening of July 15, Jason persuaded him to go jogging. This was inconvenient for the Commissioner, because his clandestine contact with Europol was scheduled for midnight, but he agreed anyway. After the jog, they sat together in the bistro. Jason was restless—*he wants to tell me something*, Storm thought. And he felt the same way; he would have liked to talk to his friend about the strange vanishing message he'd received in the simulation and about the attack at the tanks. But that was impossible.

Over a beer there was an informal opportunity to talk to Jason about this feeling of the unreal.

"Maybe the CE-4s are right," Storm mused.

The physicist shook his head. "The CE-4s? No way. That hurts my intelligence. They're too stupid. If they happen to be right despite their stupidity, I'll return all my diplomas."

"What do you mean, 'if'? Do you suspect anything?"

Nygard just stared into his beer.

The detective's natural instinct was aroused. Storm knew the physicist had his suspicions. He needed to push harder. "Come on, talk to me. You once hinted that something mysterious was going on."

Nygard sighed, took a long sip. "It doesn't taste very much like beer, does it?"

"Yeah, okay, that's fucked up, but not suspicious. Half your body, I don't know, maybe your mouth was damaged too—they grew it all out of the retort. Does that surprise you then?"

"You got a point."

"Besides, you're frozen food. What do you think the stuff is like after a hundred years?"

"Let's drink to that."

They toasted each other, two cynics among themselves.

"However"—Jason raised his index finger and looked around furtively—"talking of food, you are right. Hypothetically."

"I'm listening."

The third beer was effective, despite the bad taste.

"The measurements I told you about I now have more accurate readings. C-14 is no longer detectable in our food."

"What does that mean?"

"Well, that means that all radioactive carbon, meaning C-14, has decayed since the launch of EXODUS."

"Decayed? That means it was there to begin with?"

"C-14 is radioactive, so it decays into normal carbon. On Earth, it's constantly being created in tiny amounts by cosmic rays—energetic protons and the like—in the atmosphere. Plants, animals, all living things absorb it. When the metabolism stops, the supply is gone. The longer something is dead, the less C-14 is in it."

"Makes sense to me. How does it work here?"

"No new C-14 is produced here because the pipeline steers all the high-energy protons away from us. All C-14 was brought aboard at launch and has since slowly decayed to normal carbon C-12. There is no trace of any."

"Which means what?"

"It means we've been riding in this damned hellhole for a long time."

"You mean more than nine years?"

"My meter's not very accurate. There's a heap of lousy equipment on this rotten ship. I can barely detect a quarter of the natural amount."

"Tell me!"

"The half-life of C-14 is 5700 years."

"That's all Greek to me."

"Okay, I know you're handicapped when it comes to arithmetic—just arithmetic, I mean, of course. Here's the translation: It means we've been traveling for ten thousand years. At least."

Storm laughed as if he'd heard a bad joke. *Someone's screwing with me again,* he thought.

"You mean to say that we've been in the freezer all that time?"

"At least."

They stared at the disappearing foam of the beer, each alone with his absurd cogitations.

"There's something else," continued the physicist. "The relativistic dilation of time."

*Now comes a lecture,* Storm thought. He ordered another beer.

"Our clocks run slower compared to those on Earth."

"The time slip you don't want me to think about?"

"Exactly. We are now over 300 days younger than Earthlings, if they still exist. Earth time is always broadcast on the news. It corresponds exactly with our acceleration, so the speed is also correct. I've done the math."

"There you go."

"But only when the EXODUS is not rotating." Nygard took a long swig.

"Why?"

"It's complicated. You don't want to know. Anyway, the centrifugal force slows the rate of a watch down again. I checked it with my atomic clock. I set it up on the last duty in Level 4. And it should read two-tenths of a microsecond less than it reads."

"We're not rotating at all?"

"It means either we're not rotating, or the radio transmitter we're communicating with is rotating along with us. And since I'm pretty sure we're rotating..."

"The gyroscope!"

"You've got it. They selected me to give a course on the physics of gyroscopes. Unfortunately."

"That's so sick."

"Not so much. When you add in the missing C-14, there's a certain logic to it."

Nygard gave Storm a challenging look, silently urging him to figure it out. The inspector pretended to be thinking hard, but he wasn't getting anywhere.

"Okay, physics is not your cup of tea. Alright. So I'll repeat it slowly for you: The EXODUS is spinning, there's no doubt about it, my gyroscope proves it. So the transmitter station is spinning with it. So it's not on Earth. It must be somewhere else where it can spin in synchrony with my atomic clock. There's only one object within a few light-years that I know of that can do that."

"Our ship? The EXODUS?"

Nygard nodded his approval.

"So the transmitter is *here*, not on Earth?"

"Looks that way."

"Why should they do that?"

"Because Earth doesn't exist anymore?"

"But—"

Nygard raised his hand. "Wait, I'm not finished. Now C-14 comes into play. The vegetables, the retort meat, us, the ships—what if it's all been in orbit for ten thousand years? We were cryo-preserved in there, only nine years ago we crawled out of the tanks like life out of primeval slime.

I think the ships are an ancient deep freeze, designed to preserve the human race. At some point in the last ten thousand years, humanity became extinct, and the Earth was rendered uninhabitable or destroyed. By ourselves, by aliens perhaps, who knows?"

"But the news from Earth..."

"Fake news. Even the time slip is fake—calculated exactly how it should be, if the transmitter were on Earth. They simply forgot to take the rotation into account. At least they want to tell us that it's getting worse and worse down there. They are preparing us gently."

"Preparing for what?"

"The truth, perhaps? I bet when we arrive, they'll reveal the final disaster. And then the transmitter will go off—all dead. Only it happened a long time ago. When it became clear that no one was alive out there, an automatic countdown started. Like in the story of Clarke..." He started ruminating.

"What?"

"A space odyssey... The monolith... Who cares? The ships were activated and began their journey to a new home. Seed ships. And the irony of the story, we're not the best of the species. We were just there. We're the pathetic patched-up remnants of war heroes, our bodies are strangers to us, the beer tastes like shit, we can't even fuck properly. We're outcasts, cripples who happened to fall into the right research program at the right time. Useful idiots. And that's why we were made to believe that the Earth still exists, that we are a vanguard for the great exodus of humanity. It's more bearable than the truth."

"Who?"

"What, who?"

"Who is making us believe that?"

"I don't know. My gut tells me it's Ahlgrim, the motherfucker. I credit him with everything bad."

"So if it is him, we're also looking at everybody who didn't come out of those tanks. Molander, Thelen, Padmé."

"Maybe. Maybe they're all in on it."

"Maybe they're all softies."

"It would fit."

"Is this Jason's conspiracy theory?"

Nygard nodded sadly. "I suppose so. But at least the official version has been disproved. Physics is relentless."

Ten thousand years ... the long thumbs of officers came to Storm. A century wasn't long enough for evolution to adapt the human thumb to smartphones (how could he have believed that?), but ten thousand years perhaps. Were the reconstructed ones really something like archaic animals in a modern Jurassic Park, guarded by the long-fingered?

Conspiracy theories cannot be falsified, the physicist had said a long time ago. So they are true. But that's not what he meant. Storm thought he understood it, and now it was gone. He had drunk too much. He nodded at the clock and was terrified. It was ten minutes to midnight.

He muttered, "I'm late. I have to..." Jason tried to hold him back, reached out his hand—and flashes of light sparkled as a galactic proton shower flooded Storm's retinas. His movements froze, paralyzed for seconds. His mind went blank. *Of all times*, he thought, more angry than worried. When it was over, he wanted to shake Jason's outstretched hand, but the physicist still sat frozen, with his mouth open to speak.

"Good evening, Commissioner Storm. I am the chief officer in the Zigmund case." The announcement sounded from the ship's loudspeakers.

Storm looked around helplessly. "Good evening," he said, vaguely in the direction of the pipeline.

"Yesterday's attack on you forced us to bring forward the GCR event."

Bring it forward? Just like that—we're sending the cosmic rays a little earlier. Fasten your seat belts and follow the flight attendants' instructions. It was so absurd.

"Who attacked me?" he asked, because he couldn't think of anything better.

"It's someone who wants to prevent the investigation of our case. We have three minutes. If that time is not enough, we will initiate a second GCR event. Please speak."

Storm drew a deep breath.

"I suspect Carol Beauclere is dead. She'd been lying in the foundation of the ceramic furnace for 20 years. She'd never been to South America. Zigmund killed her and walled her up. She was going to leave him."

"You mean the old kiln on the abandoned farm?"

"Exactly."

"We'll check that out. Do you have information from Nordica Henderson?"

"There was a meeting at company headquarters in Vienna one week before the Henderson attack: It involved a secret project to stimulate nerves with terahertz waves. Version one of the same software we use for the exercises here. Present: the head of THZ, Zigmund, an employee of EUROFORCE named Hasselborn and an unknown employee of the Ministry of Peace—about 50 years old, about five-seven tall, high voice, a birthmark on the left cheek, bald head and glasses. EUROFORCE asked Zigmund to use civilians as test subjects. There was an argument about the cost. Henderson's boss later told her not to talk to anyone about the details she overheard. Otherwise, she and he would lose their jobs, if not worse."

"I think that's enough for now. Thank you, you've been very helpful. Do not discuss this conversation with anyone. It could be dangerous."

"Dangerous how?"

"As I said, someone is trying to prevent the case from being solved. Take care of yourself. You'll be hearing from us."

"Wait, wait, wait. Why have you reopened the case—"

The loudspeaker was silent.

Jason was still frozen. The other people in the bistro were also frozen, as if a movie had been stopped.

"—after a century—" Storm's voice failed him.

Then Jason came back to life. He blinked, his head bobbing and shaking. "Damn protons," he mumbled. People started moving and talking at the other tables too, and seconds later life went on as if nothing had happened.

| | |
|---|---|
| Distance to destination | 0.628 ly |
| Time to arrival | 3 y 213 d |
| Distance from Earth | 3.615 ly |
| Speed | 0.339 c |
| Acceleration | −0.098 g |
| Slip correction | 307.60 d |

# Episode 14 Contact with the Enemy

#news flash 1.7.2220

Due to stratospheric ash, the temperature has dropped by several degrees worldwide. Thunderstorms are causing massive crop failures. Famines in Russia, USA, and China.

After the expiry of the ultimatum to return expropriated assets, China declares war on Europe.

A Korean nuclear attack destroys Calgary. Korean troops enter Canada via Alaska, allied US troops occupy the east coast. End of the Canadian–American ice war after Canada surrenders. The USA and Korea gain access to vast methane ice deposits.

© RG channel via moonbase Clarke II, sent 18.11.2216

A few days after the start of the next shift, the JIAN TOU was sighted. The EXODUS's five-meter telescope detected the fusion flame of the other ship's thruster. As the Chinese ship was slower, they were constantly catching up and would overtake it in a week's time, or to be more precise, by then they would pass at a minimum distance of about 10,000 km. This soon became the subject of general conversation.

The relative speed between the ships was 46,700 km/s. The Exodus intercepted the Chinese ship's radio traffic, but of course, it was encrypted, just like their own. The atmosphere on the EXODUS was strangely agitated and gloomy at the same time; an indefinite threat could be felt. After a few days, the commander called a meeting. No intrepid Coriolis leaps from the balcony this time; he entered the arena with his head bowed, accompanied by Thelen and Padmé.

"As you know, a week ago news came from Earth that China had declared war on the EU. We are approaching an enemy ship."

The Commissioner knew what was coming.

"The JIAN TOU is dangerous. They'll attack us. I've instructed the weapon guidance system to calculate a preventive early strike. We'll destroy them long before the closest approach."

Jason Nygard, who stood close to Storm, shook his head. "Bullshit," he murmured.

"The time slot is small, so we must act immediately. If we're lucky, we'll surprise the enemy before they attack us. This is not only our duty of honor for Europe. It concerns our own future. The habitable zone on Atlantis is small, and the New Europe you are going to build needs habitat. There is no room for people who, after centuries of isolation and ignorance, had nothing better to do than to copy everything that Western science has created. Cheaper and worse they have copied it, and in the end, they have bought up our land and our culture and are exploiting us in cold-blooded fashion."

He pointed his finger at the audience like Uncle Sam on old American recruitment posters.

"We need you! You are the founders of a new humanity. The best genes of Homo sapiens are aboard this ship, safely stored in the gene banks. Your duty is to take them to their final destination and make Atlantis the cradle of the Good, the True, and the Beautiful!"

He was so engrossed in his eulogy that he hyperventilated. After a pause for breath, he turned to the First Officer.

"Thelen, the suitcase, please."

Thelen placed a suitcase on the narrow podium and unlocked it. The process followed the archaic release pattern from the Cold War of the twentieth century, except that the suitcase had an outside display for all to see. The two of them typed in their codes, yellow and red flashed, and the four turrets were made ready for action.

"Where is the Artillery Officer on Duty?" Ahlgrim asked.

In the front row, a big massive man stood up.

"Is that him? I should know him."

Thelen nodded. "This is Lieutenant Zimmermann."

"Lieutenant, would you have the courtesy. We need the launch code."

Zimmermann did not move. He merely said, "I can't do that, Cap'n."

It took Ahlgrim a few seconds to process that.

"What did you say?"

"I can't do this, Cap'n."

Ahlgrim tilted his head. "Okay. I'm sure you'll be able to explain the reasons to me. I'm listening."

Ahlgrim was suddenly calm and composed, as he had been during the simulated mutiny.

Zimmermann cleared his throat.

"The fact is, Captain, we recently learned that almost four years ago, China declared war on the EU. But we don't know how long it has lasted or if it will last at all. If peace is restored now, the JIAN TOU is not an enemy ship anymore. And if it's not an enemy ship, we must not attack it. We don't know that. And when in doubt, I cannot enter the code. Cap'n."

"Hmm, good point, Zimmermann. But you realize they're going to attack us."

"I think not, Captain. There are reasonable persons there."

Ahlgrim laughed heartily. Folding his hands, he placed his index fingers to his lips. "You realize that your refusal to comply with martial law is treason."

"Yes, but..."

"But what?"

Zimmermann lowered his head.

Ahlgrim started pacing back and forth with his hands behind his back. Then he stepped back and looked at the conscientious objector in the same way that a disappointed father looks at a lost son.

Then he turned to Thelen and said, "Shoot this man!"

Thelen stared first at Ahlgrim, then at Zimmermann, shaking his head, which could be interpreted as refusal or incredulous astonishment at this request.

"I won't do that," he said.

Ahlgrim walked back and forth between the First Officer and Zimmermann, muttering so quietly that Storm barely understood: "Everything has to be done by yourself," he drew his weapon, took a step toward Zimmermann, put the barrel against his forehead, and pulled the trigger.

Zimmermann stared at his killer in surprise, then collapsed.

The commander stepped up to the desk, placed the weapon next to the suitcase with the red ready signs, and made an unemotional statement:

"I need the second artillery officer now."

There was movement in the back row, a lean fellow came forward. He tried to hide his nervous tremor as he stepped in front of the commander. His shot comrade lay ten feet away from him.

"Rank, name?" Ahlgrim asked.

"Lieutenant Harris, Captain."

Ahlgrim moved the weapon as if he were arranging an office desk.

"Tell me, lieutenant, do you agree with Zimmermann?"

Harris shook his head.

"Then would you be so kind as to enter the launch code?" Ahlgrim made an inviting gesture. Harris stepped to the console and nervously removed a piece of paper from his uniform. He read the code and typed it in laboriously

The final red button on the suitcase display lit up and the system began the countdown. Still 4:11:12 until the launch of the cluster bombs. These would collide with the JIAN TOU one hour and 18 minutes later. As Jason explained, a relative speed of 46,000 km per second was all it took for a single bullet to penetrate the outer shell and rupture a meter-sized melting crater. Should it hit the radiation shield, it would impact with an explosive force of almost two hundred Hiroshima bombs, triggering nuclear fusion through inertial confinement, and the shock wave would kill the crew.

Ahlgrim closed the case and put his weapon back in its holster. "Anyone else who feels like disobeying my orders may step forward now."

No one moved. Lieutenant Zimmermann lay there as if asleep.

And again, Storm had that feeling of being a minor character in a game. *The captain didn't actually shoot the lieutenant,* he thought. *Someone will push the reset button, and Zimmermann will wake up.*

\*\*\*

The Chinese ship was just a dot on the panoramic screen, even with the deep space telescope. Actually, they couldn't see it, only the glow of its thruster. Almost the entire crew had gathered in the Belt. The countdown display on the screen showed 00:35 to impact and falling. The aluminum bullets had been launched one hour and 18 minutes earlier so that they would collide with the JIAN TOU when the EXODUS was still 14,000 km from the Chinese ship.

"I understand that the early impact prevents a counter-attack, right?" Storm addressed Jason. "But if the Chinese who probably had similar plans survived the attack their own artillery fire would be much more accurate as the distance between the ships diminished."

"It won't work like that. Ahlgrim is an idiot," Nygard declared enigmatically, staring at the screen.

At 00:02 the screen flickered, and the software zoomed in on the JIAN TOU. There was a brief flash in one of the central pixels, followed by a stronger burst a fraction of a second later. And then the fusion reactor of the JIAN TOU exploded in an unspectacular glow.

Suddenly alarm. The crew was paralyzed, the horn was blaring. Then silence reigned, and everyone waited, perplexed, their gazes flicking back and forth between their fellow workers and the flat screen. The afterglow of the fragments from the explosion lasted a few more seconds before the starship disappeared from their screens forever.

Now the crew of the EXODUS could breathe again. Some clapped cautiously. Some let out a "Hey!" or another pathetically inadequate interjection. They didn't really know how to react. Each person avoided the gaze of the others and left in silence.

"What about return fire?" Storm asked. "Did the Chinese serve us an ace before they went up in smoke?" His attempt to make it sound casual failed.

"The scare is over," Jason explained. "We passed the wrecked ship a fraction of a second after the explosion of their reactor. If they had fired, we'd be dead by now."

"What about that alarm, though? The horn blaring in our ears a second ago?"

Nygard contacted the bridge, and soon he had an answer to Storm's question. "The first officer says a hydrogen tank on the JIAN TOU just missed us. The reactor's explosion sent it on a collision course." And to the bridge: "Give us the high-speed camera recording on the screen."

The faint glow of the wreckage was replaced by the grainy image of a sphere from which, like tentacles, some torn supporting struts protruded. The sequence of ten images was played back in slow motion: the sphere grew larger as the hydrogen tank approached EXODUS, then rapidly shrunk again. Nygard repeated the scene several times and stopped the slow-motion at the fifth frame, the closest approach.

"Pretty close," he muttered. "If that thing had hit us..."

"How close?"

"About a thousand kilometers."

"That's *close*? It looks like that thing there was no more than a few hundred meters away! And why is it so grainy?"

"High speed. The image sequence is ten milliseconds long. It has to be very fast. We're flying past it at 15 percent of the speed of light. And there's no Photoshop out there."

"15 percent of the speed of light—sounds like a lot."

"It's about 30,000 miles a second. Once around the Earth in one second."

"That's weird."

Storm stared at the picture on the flat screen. It looked like a herald of death. *Not a good omen*, he thought. Ahlgrim almost killed them. Then he realized something. He tilted his head and pointed at the frozen image.

"Shouldn't that be squeezed? This... What was that called in your lecture?"

"Length contraction. Well observed, Commissioner, but at this low speed, it doesn't even add up to two percent"—he calculated in his head—"At most, Michelangelo could have seen it with his naked eye."

Nygard considered serving more physics, prepared to talk, then abandoned.

"What?" demanded Storm.

"Shall I tell you something? Even Michelangelo wouldn't see anything. There is no length contraction in snapshots. A sphere always appears as a sphere, although at high speed in reality it's an ellipsoid."

"Then what's all the fuss with these Lorentz formulas?"

Nygard sighed. "It has to do with the travel time of light. From further away, it takes longer for the light to hit your eye, so you see the more distant parts of the sphere where they were earlier. This apparently stretches the object in flight direction, just sufficiently to compensate for the Lorentz contraction. If you want, I can explain it to you over a pint."

Ultimately, they stuck with the beer. Instead of discussing an explanation, they celebrated the fact that they'd survived. Perhaps Ahlgrim's decision hadn't been the worst after all.

A dark time began on the EXODUS. Fear spread—fear of Ahlgrim's terror, fear of doing or saying something wrong that could be interpreted as sabotage, fear of an order that would have to be carried out against one's better knowledge and conscience.

Oliver and Jason now held their beer evenings in private during the breaks. Even the meetings with Storm's troops suffered from the example the commander had set. No one had pressed reset, as the inspector had fantasized, and Zimmermann had not got up again. Now caution was the rule. People no longer said, "Who came up with this nonsense?" when they were given a seemingly pointless instruction, but "It'll be good for something." Even the "something" had the appeal of the subversive, and so every word was weighed before it was spoken.

The mistrust grew on the breeding ground of conspiracy theories, and for Storm there were now two: that of the CE-4 adepts and Jason's sober analysis of his physical experiments. Alien abduction or the pitiful remains of humanity?

Occam's knife came to his mind. Which was the simpler of the two hypotheses? *I should ask Jason*, he thought. But he didn't. Maybe the aliens were listening. Distrust grew like mold on everyday objects. And with the distrust came the fear of snitches and traitors. Storm felt he could trust only a few people; except for Nordica, Jason, and David, there weren't many. Even Horst Thelen, who obviously didn't agree with the inhuman decisions of the commander, seemed too timid, too obedient in retrospect.

Maybe the non-reconstructed were really just fake people, avatars of a machine intelligence. Maybe they were softies, as he once jokingly hinted to Jason in better times.

Storm's suspicions revolved more than ever around the attacker he'd escaped. Calvin, the cryo-preserved CEO of MW Medical, successor to Zigmund… Someone must have awakened him from cryostasis.

It could have been one of the officers. Or Amann, whom he had delivered to the pillory. Whenever they crossed paths, Amann's hatred was palpable. And the bastard knew about refrigeration systems. The inspector used Amann's working hours to search his cabin. He had the opportunity to do this several times, because as chief investigator he had access to the apartments. He searched unsuccessfully through cupboards and drawers, without knowing what he was actually looking for. Each time, he was more careless, and when he sneaked away, he sensed that he wanted the guy to know he was under surveillance.

And so it was. They met at the Belt. Amann blocked Storm's way.

"What are you doing in my cabin?" he snapped at the Commissioner, who was playing dumb. "Do you think I don't notice? You're a hack."

"Leave me alone, idiot." Storm wanted to keep jogging, but Amann bumped into him.

"Put out your feelers, you alien!" he yelled and punched Storm in the forehead.

Storm fought him off. A shoulder thrust and a push in the easy direction sent Amann into a somersault three meters high. He stumbled to his legs only to be immediately shifted into a parabola toward the small promenade. From there, Storm hurled him uphill to half the height of the igloo. But he wasn't done yet; Storm launched himself on the same trajectory, grabbed Amann by the neck, and pushed off toward the pipeline. In microgravity, he had an easy game with his opponent. The inspector floated around him like a dancer, always clinging to a piece of his clothes, and forced him into a fast pirouette, as he had learned in Spacedance. Amann was helpless. Like a child's spinning top he floated slowly to the ground, and the axis of rotation of the human gyroscope moved with him. *Jason was right, the ship was rotating*, Storm thought absent-mindedly.

Amann came up oblique to the ground and rolled like a potato further into a flowerbed, where he lay dazed. Some onlookers had gathered. Storm held on to the pipeline.

"What are you gawking at?" he cried. "Do you finally believe the ship is rotating?"

One thing was clear: Amann was ruled out as the assassin; he was helpless in microgravity, while Storm's attacker had acted very skillfully.

| | |
|---|---|
| Distance to destination | 0.332 ly |
| Time to arrival | 2 y 213 d |
| Distance from Earth | 3.911 ly |
| Speed | 0.252 c |
| Acceleration | −0.098 g |
| Slip correction | 324.10 d |

# Episode 15 VR.2 = VR.1

#news flash 1.7.2221
The European–British War is officially ended in the Dublin Peace Treaty.
Europe cedes Ireland to Great Britain.

Bulgaria and Romania conclude an alliance pact with Turkey.
© RG channel via moonbase Clarke II, sent 2.8.2217

Molander had the pillory turned back into a triumphal column, but the distrust and sense of threat that had emanated from Ahlgrim remained. The days went by, nothing happened. Storm passed the time jogging, reading, and drinking. There were no limits on alc-points like in his former life, yet he did not succeed in getting really drunk. A feeling of elation set in after a few glasses, then he got tired and fell asleep. The exciting intermediate phase in which the philosophical depth of knowledge extended to the sources of being, did not happen. This was very strange—just like the mental blanks and the short moments of out-of-body perception that he now sometimes had in the wake phase. Maybe they were really controlled by aliens. Aliens with a special interest in solving cold cases, even after ten thousand years. And these alien detectives were apparently on board the EXODUS. Otherwise, he would not have been able to talk to the alleged chief investigator in the Zigmund case without delay.

Storm was relieved when David invited him a few days later to help plan the next VR.2 mission. With David and the task force, the specter of brooding and mistrust was banished. Although the disturbing authenticity of the virtual missions disturbed him each time, he liked the methodical analysis of simulated events, the control over the world that they gained after repeating an exercise several times. It gave him a feeling of security.

David explained the scenario.

"Your task is to bring the container with the water purification plant down to position C. You pilot the cargo into orbit, release it and control the entry," he explained at the VR control center where they had gathered. Position C was a hollow at the foot of the waterfall.

"Then you land, put the intake pipes in the river and test the system."

Storm was amazed to see Nordica and Jason Nygard next to his team. He gave Nygard a questioning look who raised his shoulders awkwardly.

The briefing was short. David sent everyone away except Dana, because he only needed two people for the Mayflower. Storm had a brief discussion with Dana, who was supposed to act as the copilot. Then he asked David to relieve them from the annoying spacesuit procedure. He realized that he had already adjusted to the changed circumstances. In the past, he would have said without hesitation: "Please spare us this nonsense."

"I'll get you Nordica," said David. "She's eager to experience that, and we have plenty of room on the Mayflower." He told her about an update on the equipment—mini coils as a stopgap measure against stellar flares. Storm, who had already studied the briefing for this exercise, showed Nordica the function on the mablet.

They put the hoods on, first Dana, then Nodica. Storm came last. David handled his hood as if the cables were tangled.

"A bit stiff, this thing," he complained as he put the Rift on; it was pressing on the crown and temples. David tugged at it like a tailor making adjustments to a suit.

"Yes, I had to replace an array, so the fabric is still a bit clammy," he murmured.

The immersion was spontaneous. Storm felt a brief tingling sensation in his extremities as usual, and then he was in the cockpit of one of the transporters. Next to him was Dana, and Nordica sat behind them. They checked the instruments for protocol, exchanged confirmations.

"Okay, mission control, we're undocking," Storm reported. The transporter with five containers on board was released. The weak centrifugal force that had pressed them into the retaining straps was instantly gone. It was like an abrupt braking maneuver toward the ceiling, and they were force-free. Nordica suppressed a groan.

"You get used to it," reassured Storm. "Usually."

Atlantis hung before them, illuminated by the pink sun. They swung into orbit around the terminator and prepared to drop the cargo. Mountains and lakes near the shadow border were razor sharp. Dense cloud formations covered the Grampians.

The cargo glider spiraled leisurely toward the planet to land near the waterfall. There was little they could do except to correct the position of the glider if needed. But this wasn't necessary. The entry into the atmosphere went as planned.

Storm landed the Mayflower on a plateau not far from the depression with the container. The air was mild, and the hills in the east glowed in the low sun. Pink fleecy clouds flocculated above the rocky ridge. From there, a silver thread hung into the valley. Playfully, unsteadily in the fall wind, the waters plunged into the depths. *I know you*, Storm thought, *you are deceptive*. He consulted with Dana, and they decided to walk the short distance to the container. The buggies remained in the hold. Storm helped Nordica shoulder the backpack with the tools and the latest upgrade of their equipment—a superconducting miniature toroidal coil and its power supply.

After ten minutes they reached the bottom of the valley. The container stood on the bank of a natural pond that the falling water had formed over the years. Nordica made herself comfortable on a rock, while her colleagues rolled out two heavy suction hoses and let them slide into the water. The work under increased gravity was exhausting; Storm sat down to rest after the work was done.

Then David called in. "What's the matter? You're not finished yet. The hoses have to go deeper."

Disgruntled, the inspector stepped into the water. He pulled on the hose, while Dana stayed on the shore and pushed the chunky thing toward him. At the deepest point, the water reached Storm's waist. They were just about to position the second hose when a lightning bolt lit up the steep rockface. They paused, listening to the sound of the waterfall. Atmospheric disturbances came out of the headset.

"Attention, we're aborting." David's voice came through the static. "A flare. We're detecting hard X-rays. The proton tempest will reach you in about three minutes."

"We can make it to the Mayflower by then."

"Negative. You can't stay outside now. Stay behind the container and activate the supra-coils. You have plenty of time."

They sought shelter in the shadow of the water treatment plant. Nordica stood uncertainly in front of her equipment, so Storm helped her connect the ring coil to the power supply and fix it to the support frame. He lifted the scaffold, which weighed about ten kilos, onto her shoulders and attached it to the risers of her uniform. The loop hung not even half a meter above her head.

She groaned under the heavy load and searched for the switch.

"No!" cried Storm—too late. Six Tesla are a strong magnetic field. You should be a few meters away from any ferromagnetic material when you activate the coil. Nordica was pulled back when the magnet attached itself to the wall of the container. She hung in the risers and kicked with her legs. In two paces Dana was with her and broke the circuit, causing the helpless woman to fall to the ground.

Dana looked at Storm. Storm looked at Nordica, who sat with her mouth open like an abandoned putto, the slipped halo above her head. Then they all burst out laughing at the same time.

David's voice over the radio interrupted them. "Yeah, that was funny. We laughed too. But you should get ready. The particle wave is approaching. The flare is already dying down. I'll let you know when it's time."

So they prepared themselves. Storm and Dana checked each other's risers and the position of the magnetic coils above their heads. Then they took care of Nordica. They did not have to wait long for the light curtains. They were ghosting across the sky even at twilight, when the particles of the flare entered the atmosphere.

The three set up at a respectable distance from the container without leaving its shadow. Then Storm gave the command to activate the magnetic fields. At first, nothing happened, but then, as the light veils of the aurora became brighter, as fantastic colors scurried across the sky as if thrown by a cosmic painter, the rings above their heads began to glow in synchrony with a swell of an outlandish music of the spheres. The dipole fields of the super-conducting coils deflected the charged particles away from them like a screen against the rain. But those particles that came too close to the center of the ring ionized the air molecules above them, and the bright plasma spiraled in all directions.

"You can go back to the shuttle now," David finally instructed. "I would rather leave you in the shadow of the X-rays, but the mini coils protect only up to 20 MeV. Hurry, you'll be safe in the Mayflower!"

They rushed back to the shuttle. If they had walked more leisurely, they could have been mistaken for the Three Wise Men with their radiating head coils.

"Mission completed," Storm reported from aboard the Mayflower, which took off with roaring engines. In a wide loop, he steered the shuttle through the northern lights on the night side of Atlantis back to the EXODUS. They adapted to the rotation at the last moment and locked into the anchorage bay. A jolt went through the cockpit as the centrifugal force took hold.

David brought the crew back from the simulation. Ants running on their arms, the feeling of being trapped in a spacesuit that hindered movement and breathing disappeared, even though they were not wearing one, and the Rift showed only noise. Storm fumbled for the glasses—now he could feel them again—and took them off. Behind him stood David, who put his hands on his shoulders. He wanted to take off the hood, but David slowed him down. Dana and Nordica were already free of their equipment and gathered curiously around him.

"Hey, what's going on?"

"Don't worry, everything is okay," David reassured him. Gently he removed Storm's hood, which was stubbornly stuck. He handed it over to Nygard, who silently examined it and shook his head again and again.

"What..."

David raised his hand, interrupting him. "How was the exercise?" he asked.

"Well, not quite normal. What was that about the flare?"

David grinned, but his eyes remained cold. "We wanted to give you an experience you'll surely have often on Atlantis..." As he said that, his grin gave way to a more thoughtful expression.

"Are we still alive?" Storm asked. "Theoretically, I mean."

"We are. You got three hundred millisieverts, no cause for concern."

"Then why are you looking at me like that?"

David hesitated. "Did you notice anything strange? Was anything different from usual?"

The Commissioner pondered, looked at Dana for help. Everyone was watching him as if he were one of the guinea pigs in Zigmund's lab.

"Was the simulation as usual?" David asked. "Failures, blackouts, paresthesias?"

"Nope, everything was fine. Did you notice anything?" Storm addressed this question to Dana and Nordica, who both shook their heads.

"This is very interesting, though."

That was Jason Nygard, who twisted and turned the inspector's VR hood between his fingers and laboriously pulled out a wire mesh, precisely adapted to the hood.

"This... is a Faraday cage. An aluminum hat!" He giggled hysterically. "Those were once bicycle spokes, by the way. I recognize them. The wires shield electrical fields, so the terahertz radiation can't get through. So VR.2 can't work."

He held the wire mesh like Hamlet held the skull.

Storm raised his hand. "Wait, it can't be. I was in VR.2, I was definitely in the Mayflower and at the waterfall. Nordica, you saw me, say something!"

Nordica nodded.

"Whether she saw you is not important. Whether you were inside, that is the question here."

"I'm sure I was. I've just been taken out by David. You were there!"

"Oh, crazy spite." Nygard cleared his throat and nodded. "Sorry for citing Hamlet. Now, look, I'll try some Aristotelian logic: If Oliver was in VR.2, but Oliver's helmet doesn't work here because the terahertz rays can't get through, then the simulation is not generated via this helmet."

The realization fell like a heavy weight. It covered up all hope, and all sounds died away. Even the centrifugal force seemed to have disappeared. Nordica was the first to break the silence.

"Then the helmets are also simulated," she whispered.

Afterward they discussed what they should do. Five people now shared a disturbing revelation: David, Jason, Nordica, Dana, and Storm himself. Nordica had made it clear: if the helmets were simulated, then perhaps everything was simulated.

Then perhaps their glorious starship was also virtual.

David shared the story of the Faraday cage with them. He had been with crew 7 from the beginning; six months after the launch of EXODUS he had been released from the tank. Strangely enough, there were no reconstructed people who had witnessed the launch themselves. Only the leading officers were there from the beginning. It was also strange that there was no information about the other crews lying in hibernation—they seemed to work completely autonomously for themselves. He had often asked Thelen, Kvalheim, Padmé, and Lilly about details of how ten or fewer of them had managed the launch, or how the shift changeover had taken place. The answers had been strangely evasive.

When, after two shifts as an IT expert, he was appointed VR.2 instructor and gained insight into terahertz technology, he became suspicious. Jason Nygard's critical remarks led to the planning of the experiment—a Faraday cage designed to render the VR.2 helmets ineffective. A classic falsification experiment. The theft of metal foils was problematic because the material on board was controlled and recycled with milligram precision. The gyroscope was ideal as a source, since destruction would raise suspicion of the CE-4. The reference to the queen of hearts as an allegory was intended to reinforce this. David had made a self-experiment with the Faraday cage. The result had shaken him, although he had suspected from the beginning that something was wrong on the EXODUS. For a long time, he had hesitated to let someone in on it. It seemed too dangerous to him, too implausible, too unpredictable the consequences—until now, when the situation on board became unbearable.

Storm was disturbed. He had fallen for David's trick and in his report had handed over two poor guys to Ahlgrim's despotism. The alternative was even more unfortunate: if he had been more precise, he might have come across David, and then he would have been faced with a classic dilemma.

Jason was determined to make their findings known during a public lecture, but the others were able to dissuade him from doing so. The claim alone would only serve as a new conspiracy theory. One could be executed. Jason's objection that an execution was only virtual if the EXODUS was virtual, no one could counter. It would be an exit, but to where? A leap into another virtual world? An awakening on the vivisection table of aliens or on an earth that has been inanimate for thousands of years? Where were they really?

In the end, they did not undertake anything.

On the last day of their shift, as they prepared for the tanks, Dana said goodbye to Storm.

"I know how to get out of here," she whispered, embracing him. "I'll do it like Truman Burbank—you know, the old movie. In case I don't see you... good afternoon, good evening, and good night."

| | |
|---|---|
| Distance to destination | 0.126 ly |
| Time to arrival | 1 y 213 d |
| Distance from Earth | 4.117 ly |
| Speed | 0.157 c |
| Acceleration | −0.098 g |
| Slip correction | 332.10 d |

# Episode 16 Mutiny

#news flash 1.7.2222
The global temperature continues to fall due to stratospheric ash. The little ice age has made southern Europe and Arabia habitable again. Many refugees return.
European ROBOFORCE troops stop the Turkish army near Vienna.
The interstellar ship MATILDA of the Australian-New Zealand Federation takes off for Atlantis.
© RG channel via moonbase Clarke II, sent 18.5.2218

When they awoke from cryosleep this time, they were almost there. Proxima was now clearly visible to the naked eye, Alpha Centauri shining as a star of magnitude minus six. However, the 0.13 light-years they still had in front of them corresponded to many thousand times the distance between Earth and the Sun. The MATILDA was the prime topic of discussion. When would she arrive? What would she carry with her? Would she be a danger to the colony? Would Atlantis have to be divided? And how?

In the new shift, Dana was no longer among them. Horst Thelen announced that Dana Winthorpe had died in cryosleep due to a tank malfunction. Her body had been recycled by Crew 8 according to regulations.

The inspector did not believe this. She had had a plan. In the on-board archive, under Truman Burbank, he found an old film, The Truman Show. It was the story of an insurance broker who since his birth had been the leading actor in a permanent reality TV show without knowing it. An audience of millions enjoyed his normal, modest, externally controlled studio life, which he assumed was real. When Truman suspects the fraud, he tries to escape, but the TV team constantly invents new obstacles. In the end, he sails into the unknown in a boat and breaks through the horizon, which is nothing more than a studio wall painted with sky and clouds.

When Storm learned that one of the numerous landing modules had been so badly damaged by a meteorite hit that it had been disposed of, he knew what had happened. Somehow Dana had managed not to get into the tank. She had commandeered a landing capsule and made her way to the studio wall—somewhere far out in space.

Storm's friends were worried. The morale on board was low. Rumors were spreading. The conspiracy theories were no longer limited to the CE-4; some people thought that the regular news arriving from Earth was fake news. Or one heard from acquaintances that another unnamed friend knew someone who knew someone who had arguments that the officers were aliens. The trimmed eyebrows were there to cover the scars where the insect antennae of the aliens had been removed. Storm's squad was also suspect. A task force that supposedly rehearsed emergency operations behind closed doors was suspicious in itself. The successful repair of the spider could not dispel the suspicion. Perhaps that was also faked. Paradoxically, this brought them close to the Commissioner's own ideas.

Which of the many relative truths that circulated on board could be trusted? Was Storm's suspicion also a conspiracy theory? The boundaries of reality dissolved.

Soon there was passive resistance. A defect in the irrigation of the gardens led to the loss of an algae culture. Storm was commissioned to investigate, but his ambition to find a culprit was as weak as could be. So, after consulting Nygard, the case ended with the investigator's finding that an electronic valve control system had failed, which may have been true.

The crew stayed away from the physicist's lectures; everyone happened to be on break or sick the same day. When subversive slogans appeared on the art wall, Ahlgrim reacted. The time-out was lifted; from now on everyone was under control everywhere and at all times.

The situation reminded Storm of the epoch of the Great Confusion, during the twenty-first century. The truth was relative, his own perceptions doubtful. Once again the unwritten social contract dissolved, its virtual shreds dispersed in the maelstrom of total control. Everyone distrusted everyone. The more he considered it, the more likely it seemed to him that informers and spies were among them. Even behind people who were unconditionally trusted, there could be unknown actors. If everything was a game, it was enough to use the appropriate helmet.

One day a military patrol picked up David Müller. They were in the middle of a briefing for a simulation that would test an alternative procedure for landing on Atlantis. The inspector wanted to know why David was arrested: it was an order from the captain.

The reason for the arrest was obvious, and in the evening Storm consulted with Nordica. She and Jason were the only ones he still trusted. If news of the Faraday cage had made it to Ahlgrim, someone in David's troop must have betrayed him. No one else knew about it.

"Even if you're right, Ahlgrim can't hurt David," Nordica reassured him. "He's too important for VR.2."

She turned out to be wrong. After two days of uncertainty—David remained in custody—Storm's troops were summoned to Ahlgrim.

"Sit down," he ordered. Thelen, Padmé, and de Vries were already sitting at the assembly table in the mess hall, where the Faraday cage lay.

"Nothing is so finely woven that it doesn't come to the light of the sun," Ahlgrim declaimed as he played with the wire mesh.

There was no response. Ahlgrim sighed.

"I won't even ask if you knew about it. That is evident. You were all there when the Commissioner had it on."

"And one of us told you," mumbled Storm. He looked around, but couldn't detect any guilty reactions.

"I have my sources. They are also necessary with this mess of a crew. Yes, it's difficult. We've been in this tin can for a damn long time, it pisses me off too. It's not funny here, despite all the fun we give you. But everything has its limits. The caricature on the art wall, that drooling thing with feelers and a curly tail, do you think I didn't recognize myself? Of all the crews, you are the worst."

"Now what does that have to do with the Faraday cage?" Nordica asked.

"It has to do with the fact that insurrection is fundamentally unacceptable. But anyway, let's talk about the facts. There are two criminal acts at hand. The first is damage to war material. That is sabotage."

Ahlgrim paused to emphasize the importance of the crime.

"Second, and now I come to you: You have attempted to uncover a military secret by unauthorized means. It is a secret weapon that is at stake. Of course, you don't know anything about it, because you're barking up the wrong tree. But the fact remains: you're guilty of conspiracy and treason."

Nordica suppressed a laugh. "You're right, this is really the wrong tree." She spread her arms and looked around.

"I'm glad you're still having fun," replied Ahlgrim. "You were like that before. You loose types make me sick. You are a security risk. With Müller, the matter is clear. With the others, I'm still thinking about it."

"Who do you mean?" asked Thelen.

"I'm speaking of Storm and Henderson, of course." Ahlgrim pushed a paper across the table. "This is Müller's confession and the grounds for his conviction."

Thelen frowned and leafed through the file. Padmé nodded reassuringly.

"As for Müller," Ahlgrim continued, "this was a deliberately planned action, designed to undermine the morale of the crew. And as we see, it worked, discipline is down. As unfortunate as it is, we are losing a good man, but he is endangering the entire mission."

As if in thought, he propped his elbows on the table and put his outstretched index fingers to his lips.

"For good reasons, I am not a humanist, but a pragmatist," he continued. "A French philosopher once said that the furor of compassion connects Rousseau directly to Robespierre. I would like to add: the furor of humanitarian welfare connects Marx to Stalin, Jesus to the Inquisition, Mao to the Cultural Revolution. Humanism is fundamentally destructive. My judgment is sound."

De Vries pleaded: "Under Article 7 of the martial law, we have something to say about this."

Padmé nodded approval. Ahlgrim acknowledged her with a dismissive gesture.

"Müller remains in solitary confinement until the verdict is pronounced. The others may leave."

Days later, Storm requested a visitation permit, which was granted. David Müller was surprisingly calm.

"It's good to see you. Did the aliens agree to your visit?"

"Are you being treated well?"

"Can't complain."

"Who was it? Who told him?"

Müller shrugged. "Maybe nobody. I've blocked the VidCams, but if I'm right, then they know everything anyway."

"You mean—?"

"Ahlgrim was babbling something about a secret weapon. That's bullshit. Who's he gonna use it against? We're going into virgin territory."

The inspector nodded.

"I think the whole thing's a conspiracy. No aliens, Ahlgrim is too human for that."

"Human?" Storm laughed dryly.

"In the sense of disgusting."

David bent over, speaking empathically: "Think about it: The permanent ones—Ahlgrim, Thelen, and the others up there—are all not coming out of the tanks. We have no information about the other crews who are supposedly hibernating. Do the tanks exist, or are they made where VR.2 is made? Maybe there aren't even—"

David flickered. Storm's mind went blank. This time the episode lasted unusually long. His arms and legs tingled as if under electricity, and he thought he was floating. When he could see again, David was gone. They had taken him. He looked at the clock: the visit was over. No one stopped him when he left the holding cell.

The verdict was delayed. For days nothing happened. Groups discussed the case in the arena and on the promenades. The mood was on the brink. Storm consulted with his people. He planned to rescue David and, if necessary, arrest Ahlgrim. The plan was dangerous, even if they had no mole in the group.

Jason had avoided him for days, but now he approached him. He'd convinced friends and colleagues that Ahlgrim's terror must be ended. They would act. Storm persuaded him to wait for the verdict. If it was lenient, they would do nothing.

And then Ahlgrim announced the verdict in the arena: Pillory sentence for Oliver Storm and Nordica Henderson for aiding and abetting high treason, and death sentence for David Müller for sabotage, conspiracy, and high treason.

When the guards moved to arrest Oliver and Nordica, who were standing in the back rows, Storm's troops intervened. Two of them tied up the guards, while two other armed men rushed forward and kept Ahlgrim and Thelen in check. More rebels came from entrance A and kept the group from the other side in check. Padmé was not present. Nygard's confidants broke away from the crowd and flocked like a shield around the leaders of the rebellion.

Storm stepped forward.

"Captain, you are under arrest. Your inappropriate actions undermine the morale of the troops, you endanger the mission. I'm temporarily assuming command of this ship. You will have an opportunity to defend yourself before a select committee of this crew."

Ahlgrim was less startled than amused.

"This reeks of a minor mutiny, my dear old friend. You've always been uncomfortable, so I tried to educate you. It was a challenge to show you where the power is. Thelen, explain to these people what awaits them unless they lay down their arms immediately."

Thelen, who stood next to the commander, took the floor.

"I cannot agree with you, captain. This is not mutiny. Your irresponsible actions have triggered Article 13 of the European Martial Law Convention."

Padmé appeared. She floated down from the terrace behind the arena in a bold Coriolis arch and landed next to the column of shame. Like a Greek goddess she stood there, and the murmuring in the arena died down.

"Thelen is right," she proclaimed. "According to Article 13, the commander of a warship, if he is incapable of action or if his actions put the mission in serious danger, loses the supreme command. The second-highest officer takes over."

"Have you all gone mad? I command, and you obey!" cried Ahlgrim. The crowd remained silent.

"No one has the right to obey," murmured Jason, who stood beside Storm.

Thelen took the floor.

"Okay, you heard it. I am taking command of this ship under Article 13. This is not a mutiny, but a regular procedure. Guards, seize him!"

Everything had happened so fast that the guards didn't quite know what to do. After a short hesitation, they followed the order. Ahlgrim was disarmed and taken away. His followers in the small flock could be recognized by their unsettled expressions. There were not many of them, and they did nothing.

Storm's team gathered at the VR headquarters. Ahlgrim, tied to a chair, fidgeted, cursed, and screamed as David Müller put the VR.2 helmet on him.

"You will regret that! I will bring you to court martial! The whole rotten bunch! Thelen, the fucking softy!"

"You'll have to wait till we get there. But we'll cut that short, we'll send you right now," David grinned.

"What? No, damn it, you can't! Without me, everything will collapse. They need me on Earth!" He shook the shackles and tried to throw off his helmet.

"On Earth? Why?" asked Storm.

"Because—you don't understand."

The inspector had had enough. "Okay, we're flying," he said. "As we discussed—Tom is coming with me, Cat and Steve in the second longboat."

They put their helmets on.

"Hold on!" Ahlgrim wriggled in his shackles. "I need a doctor. I'm not feeling well. I'm gonna die down there. Get me de Vries!"

David looked at the raving man critically. "I think you're not well because you can't bear to see others act against your will."

"At least give me an exit point. Fucking hell! I'll die down there, then the ship will be destroyed!"

"Bullshit."

"If you die, we'll have fun," growled Tom. "Can't you let him rot down there?" This to David.

"When he bites the dust, he automatically ends up with me. Then I'll send him back down again—Groundhog Day. He can't get away on his own, he doesn't know the exit point. But he won't die. There are no evil aliens down there. At least for the time being."

Storm gave his okay, and they put on the Rifts and the immersion started. Ants running on their arms, flickering, white noise, and they were in the shuttle. Ahlgrim was handcuffed but screamed and raved so much that Tom had to keep him quiet with duct tape.

Many times they had practiced landing, the procedure was familiar to them. They released themselves from the rotating Exodus. Suddenly they were weightless, it was like letting go as the Mayflower drifted tangentially away from the mothership. The Milky Way weaved a silver ribbon into the infinite emptiness of space. Alpha Centauri's blue light was so bright that it cast shadows on the longboat. Familiar constellations, star groups, globular clusters, stars everywhere. Atlantis hung in front of them. They swung into orbit, and there Proxima rose radiantly above the horizon, four times larger than a sunrise on the lost Earth. Even though it was a simulation, Storm felt a sense of elation every time Atlantis came into view. It was the hope for a new life in a new world, even if it was a virtual one.

The Mayflower landed on the high plateau of Cape Cod in a crosswind. The second boat remained in orbit as a reserve. The rover quickly brought them down to the valley. The containers were now placed at a safe distance from the treacherous creek. Tom parked the rover next to the weather station. Storm checked the surroundings. He judged the weather situation to be harmless: The brook gurgled peacefully, the waterfall on the western mountain cliff was a glittering silver band, above the ridge hung a small cloud in the pink sky. Proxima stood higher this time, the shadows were shorter. The air was fresh and cool.

He tore the handcuffed man from his seat and removed the tape that had prevented him from screaming and cursing. Ahlgrim seemed to have given in to his fate. Storm handed him over to Tom who was to lead him to the container where he would spend his time as an interstellar castaway until— yes, until Horst Thelen brought him out of exile.

"I'll take a look at the settlement from the hill," the inspector said. "In the meantime, explain what he needs to know. Food rations, drinking water, photovoltaics, medicines..." He turned to Ahlgrim. "There are seeds in the depot, even a manual. You can work as—" he actually wanted to say farmer, but then the abandoned farm from the past flashed up almost tangible before his eyes; the dead vineyards, the burnt fields, Carol, Zigmund, the bomb attack, and he said: "—work as a peasant when you get bored."

Storm brusquely turned around and climbed the hill on which the container colony was built. From the hilltop, he had a view of the settlement. More than fifty containers were grouped around the energy center with the solar panels—living space for 600 people. This was their Shangri-La, a valley protected from the windstorms. The Grampians shone like a theater backdrop: Here and there orange-red jagged edges, pinnacles, and cracks against a black background in the grazing light. The sky was clear except for the small cloud above the waterfall, which fell as a silvery ribbon over hundreds of meters into the valley. The climate was constant and mild. On the eastern slope and along the creek the settlers would cultivate fields. A good place for a departure into the future.

"The captain wants to see you," said Tom when they met in front of the container. "He is very calm, playing the serene one again."

They trudged silently down the hill. *If I only knew the exit point, everything would disappear*, thought Storm. First Tom, then the others. In the end, it would be his turn, returning to the real world, to nothingness perhaps. Not a bad solution. If the EXODUS was an illusion, why not himself? Maybe he was a Boltzmann Brain, a random structure in the thermodynamic disorder of the universe, dreaming.

Ahlgrim sat there confident and calm; maybe he thought he could go on playing boss, or maybe he had resigned himself to his fate.

"You wanted to see me?" It was supposed to sound like a statement, but somehow the sentence became a question. The emperor of Shangri-La held audience.

"You're a good cop, Oliver Storm. Your old case still haunts your mind. But you think one-dimensional. Always have. You think you know the truth in your cold case, maybe you're close. But it's no use—this is a one-way street. There is no turning back. Free yourself from yesterday."

"Is that all?"

"Fate likes you. You're damn lucky to have survived two attacks."

"Two? How do you know?"

The prisoner smiled. "The captain knows everything."

Storm charged at Ahlgrim, clobbering him with his fists. The chair tipped over and Storm continued to strike, blinded by rage and despair. Tom tore him away from his victim. Otherwise, he would have knocked Ahlgrim unconscious.

"Damn—he—he's behind this, he sent Calvin on his way!" This to Tom, who stared without understanding.

"Who?"

No one else knew about the attack. Tom probably thought he was a maniac.

"Aw, forget it."

Slowly, he calmed down. It was just an illusion. How could he let himself go like that? He stepped back, massaging his aching fists. Ahlgrim let himself be tied up with duct tape. Tom unlocked the handcuffs so he could easily free himself, but by then they would already be gone with the rover. The EXODUS had emerged from the radio shadow of the planet. Storm reported the completion of the mission, and they returned to the ship with engines blazing.

The next morning a call came in from David. He wanted to show Storm something.

"I'll send you down to see our former captain for a moment. You'll like it."

The inspector put on his helmet and goggles, and the immersion started. The surroundings flickered, static, blackout, and he found himself at the bottom of the Shangri-La valley. Now that he knew that he was just moving from one virtual reality to another, VR.2 seemed even more realistic.

The rivulet was gurgling peacefully, a light breeze was blowing, everything was quiet. Proxima stood low, its lower edge touching the ridge. Ahlgrim was nowhere. Storm strolled between the containers at the foot of the hill. It seemed to him as if they stood closer together. Suddenly one of the containers moved, it was a barely perceptible glide, it seemed to slide down the slope to the water without any hurry. Something glittered at the edges. Storm came closer to see what was going on. The container seemed to be gliding on a layer of slime that gleamed at the edges of the base plate. He bent down and saw the snails. Hundreds, thousands of slugs had crawled out of the floor under the container and pushed it down the slope millimeter by millimeter.

"What the fuck—!"

Storm's gaze followed the course of the river, and in the distance, he saw several containers standing, or rather in imperceptible motion on a leisurely walk to the lake into which the stream flowed.

Then the door of the container flew open and Ahlgrim rushed out, sprinting toward him. Storm instinctively ducked before the man, who simply ran through him and fled toward the top of the hill.

"Don't panic, it's just a simulation," Storm shouted after him. But Ahlgrim.2 could not hear him in his world personalized by David Müller with mischievous jokes.

David brought him back from the short trip. He grinned.

"I've named those slugs *Limax shangrilasius Müller*. I wanted to see if Ahlgrim could get along with primitive aliens. If he keeps running away from the natives, the entire settlement will be gone in two or three days."

Storm shook his head. He didn't understand how someone in their situation could devise such games. A small satisfaction after Ahlgrim's malice, a student's prank.

David raised his shoulders when he noticed the inspector's reaction. "I want to put him to the test. You can save the settlement, it's very simple."

Storm pondered. "You could reduce the contact surface of the module, then the slugs might not be able to carry the weight anymore?"

"Good boy! It's enough to wedge the container under on one side, ten or twenty centimeters are enough, then it just stands on one edge, and the screws are powerless."

*He won't make it*, thought Storm. A double satisfaction for David.

"When will you get him out?"

"Thelen must decide that. If he doesn't order it, I'll let Ahlgrim stew for another two or three days. Until the settlement is gone."

"He can stay there till we get there, if you like. And then we'll see what happens anyway. The Game Master out there"—Storm vaguely pointed in the direction of an unknown reality and raised his voice—"will end the game."

But he was mistaken.

Two days later Ahlgrim.2 disappeared and Ahlgrim.1 was dead. Thelen called a meeting in the arena. His speech was short—he told the crew that Commander Ahlgrim had died of a heart failure in the holding cell. The grief was limited.

David was distraught, he was not known as such. He played Nordica, Jason, and Oliver the VidCam recordings that showed both Ahlgrims; a man threatened by the natives, more annoyed than panicked, switching from one living module to the next when *Limax shangrilasius* began its work. Next to him was a dreamer sitting in the VR center whose limbs spastically twitched when his avatar once again jumped up and ran away from the snails.

Just before he disappeared, Ahlgrim.2 became uneasy. He tried in vain to activate the directional antenna, obviously to contact the EXODUS. Seconds before he disappeared, the dreamer twitched violently, like in an epileptic seizure, and his Avatar did the same. And then the cameras showed how Ahlgrim.2 faded, became translucent, and dissolved like early mist in the sun. Ahlgrim.1 hung motionless in his chair. A few minutes later de Vries arrived. He recorded death by heart failure.

| | |
|---|---|
| Distance to destination | 0.018 ly |
| Time to arrival | 213 d |
| Distance from Earth | 4.225 ly |
| Speed | 0.057 c |
| Acceleration | −0.098 g |
| Slip correction | 334.40 d |

# Episode17 Arrival

#news flash 1.7.2223

The inner-Arab conflict between Sunnis, Shiites, and China is settled in the Beijing Treaty: Saudi Arabia cedes to China the territory north of the 25th parallel and a coastal strip with access to Jeddah. China guarantees the protection of the northern border and the Persian Gulf, grants freedom of religion in its territory, and opens the borders to returnees.

Chinese forces are advancing to the Seine. Russian medium-range missiles destroy all six European fusion power plants.

© RG channel via moonbase Clarke II, sent 9.4.2219

With Molander, a new spirit emerged. When Crew 7 was awakened from cryosleep for the next shift, time-out was reinstated. The ubiquitous mistrust subsided, the mood brightened—not only because of the new commander, who radiated a calm, self-evident authority but also because they were almost at the end of their journey.

The sun of Atlantis shone as bright as Sirius, the brightest star on Earth's sky, but in a warm and friendly orange. Appearances are deceptive; their new home sun was prone to outbursts of anger. Occasionally it hurled gigantic protuberances into space, a flood of fast particles and hard X-rays that would rain down on Atlantis, the planet nearest to the star. The strong magnetic field protected the atmosphere but was insufficient during heavy stellar storms.

Alpha Centauri, four hundred times brighter than seen from Earth, was still a dot. From a distance of 0.2 light-years, the double star could not be resolved with the naked eye. Only the telescope showed the twins at a distance of two minutes of arc. The celestial constellations had hardly changed, only Sirius, the brightest star after Alpha Centauri in her new celestial tent, had moved up to Orion's shoulder, perhaps to stop the restless Betelgeuse from breaking out as a supernova. Sol, the third-bright star after Sirius still stood in Cassiopeia, at the end of the familiar "W" of the reclining queen, close to her daughter Andromeda and her savior Perseus.

Shangri-La would soon rise from VR.2 to reality. An expectation, a controlled excitement electrified the crew, and it calmed the CE-4s, who were reluctant to make a fuss about aliens. This did not evoke any euphoria in Oliver Storm and his friends; they knew that there was something fishy going on. The ineffective Faraday cage, the strange results of Nygard's physical experiments...

Were they really in a simulated environment? Then their VR.2 exercises would have been a simulation within a simulation. The imminent landing on Atlantis would be nothing more than another VR exercise. Where and when was reality?

For David and Nordica it was evident that the crew was moving in a VR scenario that was indistinguishable from real. Minor errors, such as the stale taste of food and drink, blackouts or out-of-body perception were due to the software and not to the pretended high-energy particles of the Galactic Cosmic Rays.

Jason, for his part, had already put on Occam's knife and found that when they were in a VR scenario, the physical paradoxes he had observed found a much simpler explanation than a humanity that had been extinct for ten thousand years and a ship that had been drifting in space ever since: they were in an almost perfect simulation, but the software developers had made a few small, crucial mistakes. *We must examine the tanks, maybe the real crew is there*, thought Storm absent-mindedly, until he remembered that it was nonsense.

"I know these university guys," growled Jason as the four of them sat together in the operations center and consulted. "Physics hardly ever occurs in computer science studies. How are they supposed to know such self-evident things as the radiocarbon method or relativistic time dilation? Now I also understand why there is no plasma in the funnel. They stole Poul Anderson's idea instead of calculating. It can't work at all. The magnetic funnel would get clogged at our high speed."

"Only paranoiacs notice such mistakes," Nordica provoked Jason, who did not find it funny.

"I wanted to say that good haptics are sufficient," she placated. "That's all a normal person needs in a computer game. Essentially, the software uses ray tracing. It knows from the representation of the virtual space which parts of the actors' skin touch which object. This is transcribed into a suitable signal to the helmet, which stimulates the right neurons. Sometimes this goes wrong, then there is a strange tingling in the arms or legs, and when the system notices that it is really wrong, it switches off for a second or two. This is what they call mental blanking. A precise technical term, in a way."

Jason nodded absently. "They should have asked me to help them with the simulation software. I wouldn't have made these stupid mistakes."

"But then you wouldn't have known something was wrong," Oliver interjected.

"On the contrary, then I would have known from the start. But then I probably wouldn't be here. Or would I? That would be even worse. Ah, hell, I don't know," he concluded with a resigned gesture.

They agreed to keep their secret to themselves. The crew wouldn't believe them. Everyone would call them conspiracy theorists, and Molander would have to struggle against dissension on the Exodus. All the mistrust would flare up again.

David had kept silent for a long time. Now he said:

"We can't leave the game on our own initiative, no more than our Robinson Crusoe down on Atlantis could've gotten out of VR.2. For that, we would have to know the exit point. Or we can virtually kill ourselves like Dana."

Jason grimaced but kept silent.

"It sounds crazy, but I want to know what happens next," Oliver said. "It's a game. And every game has a goal and an end. I think it's close."

"Maybe we'll wake up in paradise," murmured Jason.

The meeting broke up. Jason suggested a beer, but Oliver put him off until the next day. Nordica had thrown him a suggestive gaze, which he'd interpreted correctly.

"Let's play love," she whispered on the way home. Storm silently wondered how Ahlgrim had known about his cold case. But then he was distracted by Nordica's beauty and forgot about it. As they lay next to each other, exhausted and contented, thoughts drifting, she said into the darkness:

"It is a strategy game. Three starships and one win—we fight for Shangri-La."

Lilly Angelis visited Storm during his time-out. He sat at breakfast in the morning and waited for the remains of his dreams to fade. He was still haunted by the imagined octopus. More and more often the dream played on Atlantis, in a rural idyll between waterfall, river, and lake. Huge trees grew on the hills, flowerbeds lined the streets, and in the distance cultivated fields glowed in the sun. And there was the abandoned farm with the oven in the barn. David's snails crawled out of the meadow floor, forming tentacles that intertwined to form a huge octopus.

"I'm supposed to hand something over to you," Lilly explained, placing a sealed envelope on his desk. *Commissioner Oliver Storm, Confidential* was written on it.

"From whom...?"

"From the top," she replied and left him. He opened the envelope—*how old-fashioned*, he thought. Inside were a few A4 pages. The paper was strangely smooth like glass.

*EXODUS incident*
*Interim Report of the ESA inquiry panel*

*Secret*
  *Issued on:*

  — *The European Court of Justice*
  — *Ministry of Peace*
  — *EUROFORCE (Internal Audit)*
  — *Oliver Storm (virtual, EXODUS II)*

<u>Task description:</u>

The task of the ESA panel was the investigation of the EXODUS conspiracy, initiated by high-ranked members of the European Defense and Deterrence Agency (EDDA). The conspiracy was based on:

1. Project "Second Life," aimed at the rehabilitation of war-disabled people whose condition did not permit surgical reconstruction. Terahertz technology should allow the severely injured to enjoy an almost unimpaired virtual life. As is well known, a new defense doctrine led to the freezing of a Phase II study in 2070, as funds were diverted to the development of nuclear weapons.
2. The discovery of a habitable planet in the Proxima Centauri system made the evacuation of rulers and European elites seem realistic in the event of a nuclear attack. European fusion technology set the flight time at 13 years.

<u>The aim of the conspiracy:</u>

Project EXODUS was secretly planned to study the stress resilience during the journey—a paramount factor as the catastrophic outcome of the Mars missions had shown. The journey was to be simulated using technology and patient material of Second Life. Severely wounded warriors were to be put into an artificial coma and, after awakening, find themselves completely "reconstructed" on a starship. The conspiratorial group considered it impossible to maintain the illusion over 13 years and reduced the duration of the project to 14 months. The crew was to be told they'd experience 11 months of hibernation for every one month of service. The officers were software modules (softies) who pragmatically defined and monitored a strategy. The behavior of the test subjects (hardies), their interactions, reactions to provocation tests, mental disorders, resilience, flexibility, and stability were to provide information about the risks and chances and contingency plans for the mission. "Unfaithful pilgrims" were detected and removed from the simulation by purported death. In order to make EXODUS credible, the crew was systematically fed fake news of an Earth that was perishing in climate change and wars.

The medical care of the injured—civilians and war victims—was provided by Prof. Dr. Peter Zigmund, one of the world's renowned experts in neurosurgery and well linked to EDDA through his research company MW Medical. Medibots ensured precise surgical intervention at the scene of an accident, rapid transport by means of drones, and the transfer of the victims to a supply unit called "the tank" at MW Medical. The technical equipment was supplied by THZ SARL, with branches in Brussels, Berlin, and Vienna.

Virtual reality software was developed by the start-up RealGames GmbH, which is majority-owned by the Ministry of Peace. Due to the incompetence of the *developer team, errors in the software occurred, which some participants in the experiment were not unaware of. See the technical report for details.*

*After the failure of the democratically organized mission EXODUS I, the conspirators' hope lay with Mission II, which was militarily well-structured.*

*The exposure:*

*Suspicions against Prof. Dr. Zigmund in a series of murders led to scrutiny of MW Medical. The decisive tip-off to RealGames, the manufacturer of the simulation software, came from lieutenant Zimmermann, executed by the captain of the EXODUS II who considered him unfaithful. He was removed from the simulation and was afterward equipped with an exoskeleton in MW Medical's test lab. Zimmermann could circumvent the lab's firewall and inform the police about the malversations on the spaceship. Further examinations led to the conclusion that all the victims of the alleged serial killings were located in the virtual reality of the two ships. Another twenty victims could be assigned to the same perpetrator. The secret service had systematically suppressed suspicions prior to this point. Commissioner Storm, onboard EXODUS II, provided the decisive clue proving that Zigmund and his fellow conspirators were responsible for the bombings that disabled and mutilated the victims. It turned out that EDDA wanted civilians added to the military crew. Zigmund repeatedly delivered the victims of his crimes.*

*After the conspiracy was uncovered, the Ministry of Peace took control of the simulation. This Panel proposes to continue the illusion of EXO VR.1 for the faithful pilgrims. Revelation of the truth could cause irreversible mental damage. However, disclosure to the Storm/Nygard group is strongly advised. The psychologically unreliable ones should be withdrawn at a well-defined exit point in order not to endanger the future of the virtual colony.*

*For the panel: H. Rathenau, minister.*

The Commissioner put the document aside. He felt himself in a state of limbo between waking and dreaming, between truth and illusion, between life and death. Only slowly did elementary cognition take shape, first casually as if it were a detail, then with question marks, then with disbelief and rejection, finally with protest and anger: Oliver Storm existed. He was more than a virtual entity on a dream ship. He lived in the here and now, lived in the world of the twenty-first century, in the world of his task force, the world of Alice, of Zigmund, and his living dead.

Then came the shock—the supposed reconstruction of the victims was virtual. Oliver Storm did not exist in any familiar flesh-and-blood form. What was left of him after the bombing was in some sort of tank with a VR helmet that stimulated his brain and VR glasses that fooled him. He squeezed his eyes together in the naive hope of waking up, but it was banal as always when you squeezed your eyes together—you couldn't see anything.

Storm shared the report with his friends a few days later. Nordica and David reacted calmly. Only Jason, who had been the first to suspect the truth and always argued rationally, was terribly disturbed. Despite Occam's knife and Aristotelian logic, he could not bear the idea that their severely wounded bodies were lying in some tank on planet Earth.

| | |
|---|---|
| Distance to destination | 0.000 ly |
| Time to landing | 23 d |
| Distance from Earth | 4.243 ly |
| Speed | 0.000 c |
| Acceleration | 0.000 g |
| Slip correction | 334.50 d |

# Episode 18 Orbit

#news flash 24.2.2224

Following the surrender of the European Armed Forces, Europe will be divided among the victorious powers at the Congress of Vienna.

Calculations show that Apophis II will strike the Earth with 99% probability.

Mountain Fortress is nearing completion.

The world population has fallen to 2.5 billion.

© RG channel via moonbase Clarke II, sent 25.11.2219

Crew 7 was brought out of cryostasis one month after arrival in orbit. Molander wanted the Commissioner and his experienced team for the landing.

This was the official version. The Commissioner knew better. There was only Crew 7, and David, Nordica, and Jason also knew the truth now. They waited for contact with the real world. Lilly promised to inform him as soon as someone out there would contact her. Nothing happened, and the waiting became unbearable. Storm sometimes believed that the ESA report was also virtual.

Two weeks later, Lilly transmitted a message. The Ministry would send someone to meet him. The inspector was supposed to arrive on the bridge at eight o'clock the next morning. Apart from the emissary's avatar, only softies—Molander, Kvalheim, and Angelis—would be present.

The next morning he awoke from his confused dreams. The Octopus, Ahlgrim (or was it Zigmund?), Alice (or was it Nordica?), the CE-4, aliens in reconstruction tanks, Atlantis, Shangri-La, and the snails—everything whirled around in his head. He laboriously untangled the threads of this Gordian knot to reconstruct a fragile reality.

At ten minutes to eight, he was on the bridge. Molander stood at the control desk, Kvalheim and Lilly Angelis sat at the card table with a third person. This person turned around when Storm entered. It was Alice. An avatar who resembled Alice, actually.

"Hello," said the avatar.

The inspector kept his distance. "Hello. Kind of you to choose this avatar to ease the atmosphere. But I would have preferred to see the emissary as she is."

"Oliver, good to see you. You are looking at the Ministry envoy as she is." And after a pause, "I'm Alice Falkenberg."

His cortex first had to convince his limbic center that less than 150 years had gone by since he had seen her. There she stood, Alice as he knew her, one leg playfully stretched forward, her head slightly tilted, and when she spread her arms and said, "The Gulf Stream is still not down," he knew it was her.

"Is this really happening?" he murmured as they embraced each other. It felt strange, as if they were in a VR.2 environment, in a Forensic Time Machine that had thrown him 150 years into the past, into a perhaps not better but more authentic past.

He had a thousand questions. "Awesome"—do you look, he just bit his lip and said instead: "...to see you. How did you get into the Ministry?"

"Hanna Rathenau called me in after I uncovered the plot."

"That's a promotion. And Headless, our old task force?"

"Has been dissolved—case closed. Greetings from Julia and Horst, by the way. Everyone's fine."

They sat down with Angelis and Kvalheim, kept silent for a few seconds.

"We don't have much time," remarked the quartermaster. "Three Hardies will appear on the bridge in ten minutes at the latest."

"How's it looking down there?" asked Storm, pointing to the monitor on which Sol was beaming in the middle of Cassiopeia.

"Not so great. But there's hope. The war with Britain is over; that was no fake. The Thunberg adepts will be heard. Former Islamists care about the climate, our fusion power plant will soon be connected to the grid. The economy is pretty much down, but greenhouse gases are rising more slowly than feared."

Angelis interrupted: "Mrs. Falkenberg has come to make a proposal. It's about the return."

"The colony is being established," Alice explained. "The illusion is maintained. But there's a way to get you back from the simulation. First, the initiates."

"What are you telling me?"

"The initiates, that's you four. You, David, Jason, and Nordica. You can decide yourself. Then there are the unfaithful pilgrims, as they call the psychologically unreliable. They will be withdrawn, anyway. The ROBOFORCE army only exists in virtual reality, but there is a development project that has been massively pushed forward in the military. Zigmund's research has resulted in prototypes of exoskeletons that are mentally controlled. They would give you a body."

Storm thought it over. He and his three friends would have to take a careful look at the exoskeleton technology before making a decision on whether to exit this simulation. To help them evaluate the technology, Alice had unlocked access to Zigmund's lab for the Forensic Time Machine, so the four initiates could touch and feel the Waldos in an indirect way while remaining inside VR.2.

Molander reported: "Two Hardies approaching the bridge."

Alice got up to say goodbye.

"By the way, you remember Petrides, the second victim?"

"Yes?"

"He was Carol Beauclere's lover. That's why he had to die."

"And Zigmund? Did you put him away?"

She hesitated. "Zigmund is dead."

*Of course*, Storm figured, *for 150 years*. Then he realized that VR.1 still had him in its grip.

"Did you get him—?"

"We found him in his office. Shot to death. He was wearing a VR helmet and a Rift."

"What??"

"Remember when Ahlgrim got taken out?"

"Of course. He had an epileptic seizure."

"It wasn't a seizure. It was at the exact time Zigmund was shot."

"You mean...?"

"Yes. Ahlgrim was Zigmund's avatar."

Storm wanted to kick himself. He should've seen it. Both Zigmund and his avatar had used the same gesture—forefingers on their lips. Storm spontaneously saw other similarities: the casual-aggressive language, Ahlgrim's hint that he knew the Commissioner from before...

"Zigmund hacked Ahlgrim's softie and kept plugging into the system, so he could play the almighty every two months. When you put him on Atlantis, he couldn't get out because he didn't know the exit point. He sat helplessly in his office over the weekend, with his VR helmet on his head and his glasses on. The secretary would have found him on Monday, but the secret service was faster. They became suspicious when he didn't report in, discovered what he had been doing, and promptly liquidated him. He was a security risk."

"Why?"

"He was trying to derail the project. The simulated starships have been taking a lot of funding from his Waldos. EXODUS II was conceived hierarchically, as a counter-design to EXODUS I, which was planned as an open society. Everything was decided there democratically, and much was short-sighted or owed to convenience. When they had a bolide hit, they weren't able to repair the engine, because there was no task force, nobody was responsible. Zigmund felt confirmed by the failure. When, contrary to expectations, EXODUS II worked well, he sabotaged it by taking the military structure ad absurdum. He wanted his sadistic actions to lead to mutiny and everything to go to hell."

"But he supplied the heads himself."

"So what? Money and power..."

Storm nodded. He knew Zigmund only too well.

"And this Calvin?"

"Made up. That was Zigmund too. He overheard us trying to contact you."

*The initials*, the inspector thought. B.C.—Carol Beauclaire. That suited Zigmund.

Storm informed Nordica, David, and Jason of the option to return to reality. He reported on the Forensic Time Machine and the Zigmund prototypes. Each of the four would watch the Waldos in operation via the VR.2 environment of the Time Machine.

David prepared the equipment for Storm. Alice had set the time coordinate, for whatever reason, which only she knew: Ship time 12 July 2218. This rang a bell. Something strange had then happened, but he couldn't track it down. *She has an ulterior motive*, he said to himself as the immersion began.

There was the flickering of the Rift, the ants running on his arms, and then Storm was standing in a lift cabin. The cabin opened. On the left-hand side, a massive security door closed the way. *EUROFORCE Research. Restricted area* was written in big letters. He knew the place from his former life: on the right at the end of the corridor was Zigmund's research laboratory. He followed the directional arrow on the floor, which showed him the way. After twenty meters, the corridor should end, which is what happened. On the glazed door stood *MW Medical*. Behind it, laboratory equipment could be seen. The simulation didn't have the quality of what VR.2 usually provided. The impression was grainy, the colors gave the impression of post-coloring. Zigmund had opened this door eons ago with his personal code. Storm stood there helplessly for a while, then he pressed against the door. His hand went through the glass without resistance. That was FTM, he remembered, not much more than a 3D film, dependent on the data available in the cloud. He crossed the barrier in two steps, as if entering another world through a mirror.

He found himself in a huge hall. Nitrogen dewars, centrifuges, glove boxes on the walls, microscopes, fans, electronic components, cables on consoles in between. Along the axis of the room, there were endless rows of tables, on top of them vessels with nutrient solution in which fist-sized gray objects were floating. Wires and tubes connected the objects with pumps, flashing displays, and screens. On the tables stood various test waldos. He recognized the metal rabbit skeleton eagerly sniffing a virtual carrot.

As he penetrated deeper into the room, the waldos grew larger, including the containers and the gray objects that floated inside. A metal dog gnawed at a bone, a plastic pigeon hanging by a thread clumsily flapped one of its wings. A service technician adjusted some parameters on a tablet, and the other wing came to life. The dove's black eyes looked at him reproachfully.

Then came berths with people. A skeleton walked on a treadmill. The bones were too thin and anatomically wrong, as Storm knew from forensic training. Servo motors replaced the muscles. The skeleton went and went, doing nothing else. In the next bunk, a soldier in infantry uniform practiced close combat with a robot. The robot, also in uniform, was lifelike except for the head, which consisted of two eye-sized cameras on a cross-bar. Storm watched the turmoil from all sides; the Waldo moved quickly and smoothly, skillfully evading the enemy. Behind the fighters, half in the shade, sat a young woman in a wheelchair. She watched the fight with an alert eye, commenting enthusiastically on every success of the Waldo. Storm circled the wheelchair. Several cables sprouted from her cervical spine and wound their way to a console where a technician was working.

The inspector looked at all the berths. In some of them an automatic process took place—covering, firing, reloading in an endless loop. In other berths, patients in wheelchairs or on high-tech beds steered waldos of different types and sizes. Still others contained waldos petrified in movement, as if they had been instantly frozen.

So this was the option of a return to reality.

He had seen enough. He turned around, left MW Medical, and headed for the exit point—the lift cabin, where David would bring him back. But he hesitated before the lift. A few meters further on was the armored door marked EUROFORCE Research. This was the restricted area where Zigmund had long ago suspected that military secrets were kept, although he'd insisted he knew nothing about them. Storm slowly walked toward the door, even though he knew the power of the Time Machine should end here. After all, if the area was restricted, there would be no accessible videos or plans for the Time Machine to access. But when the conspiracy was exposed, the courts would've gotten involved, and there would've been disclosure requirements and witness statements. So maybe, just maybe...

On a trial basis, Storm touched the door's steel reinforcement. His hand penetrated without resistance. He stretched his arm, which disappeared into the steel up to his shoulder. Determined, he stepped through the door with one step. He found himself in an airlock that was closed by a second security door. The light was dim, pumps pounded, numerical codes flashed on the walls. He stepped through the second door into the restricted area.

It was bright here. At a table on the side of the lock sat a uniformed man, busy with his tablet. Storm stopped reflexively, but the soldier didn't pay attention to him. Of course, he realized, this was the Forensic Time Machine—in VR.2, but not VR.2. He saw a film showing the restricted area as it had looked on 12 July 2218, ship time.

He patted the uniformed man on the waxy shoulder. The directional arrow on the ground tried to make him turn around, but Storm just kept walking. The entrance area opened up into a hall where countless batteries of computers stood, piled up into meter-high columns. Laser beams connected the columns. Cooling systems and powerful ventilators absorbed the waste heat of the incredible computing power. Millions of petaflops were processed here, and the lasers weaved a shimmering web of ultra-fast bits between the columns. Further doors led from the walls of the square hall into adjacent rooms. EXODUS I stood at the left door. Underneath, someone had scribbled *failed* ☹ with a felt-tip pen. The room was empty except for shelves, loose cables, and hoses. At the right door was written EVAC. Inside, technicians were busy running around. Equipment was set up, cables were laid. Something was prepared here. Was it the long-planned evacuation of the elites? He did not know and it did not matter.

A similar sign with the inscription EXODUS II was attached to the middle door. On the side, a red display was flashing. Simulation active. Do not enter was written on it.

Storm childishly held his breath before entering the room. It's a film, nothing but a film, a bad, grainy one with lots of gray on top, an incomplete document, he told himself. You can't interfere with the film and it won't hurt you.

Dim light surrounded him. Machines hummed and pumps stamped softly. He stood in front of several rows of seat trough-like containers filled with a semitransparent liquid. Cables, metal rods, and pulsating hoses disappeared in the broth. The inspector moved closer to one of the containers. In it sat a male person, his shoulders half under water. His head leaned relaxed against a support, a thick full beard framed his face. He looked familiar to Storm. On his head, he wore a construction that looked amazingly similar to their VR.2 helmets. A Rift covered the area around the eyes and ears. On a blackboard it said:

*Klaus Amann*
*Engineer*

The cloudy liquid hid the lower parts of the body, but Storm didn't even have to look closely to know that there was something missing; he knew Amann's history. A bomb had blown away both his arms and legs and his pelvis. But instead of being in the reconstruction program, as they all were led to believe, he had landed here and dreamed—dreamed of being on a starship heading for Atlantis.

Storm walked along the rows—mutilated people with helmets and VR goggles everywhere. The tanks were arranged alphabetically. He recognized some names. He avoided the initial letters H, M, and N, because he wanted to remember the blonde beauty from the Arctic Ocean and his friends as he knew them.

Suddenly dull steps. He turned around. Two people in white protective clothing ran toward him, through him and turned in the next row. Curious, he followed them. They stopped in front of one of the seat tubs, handling instruments.

"Here he is," gasped one of them.

The other, loud and out of breath: "O'Brian. He's bought it."

"It was the stress—it's got him down."

Storm leaned forward, passing through the man who'd connected a device to the console—an electroencephalograph. Scales, numbers, several lines. A representation of the electrical activity in the brain.

"Not that loud," warned the first. "The acoustic shielding on the Rifts isn't so great. The guys might hear us."

They whispered now. Storm had trouble understanding them. They looked at the display.

"Storm with his shitty investigation. The control circuit raised the blood pressure instead of lowering it. Aneurysm in the midbrain. Nothing more to be done."

"Brain waves?"

"Negative."

They stood at the bathtub, helpless.

"We can't let him die now, not this close to his sentencing... it's too suspicious."

"So we'll replace him for now."

"All right. One more softie."

They took the instruments off and retreated. Storm remained at O'Brian's bathtub for a while. There wasn't much of him around. The Rift covered eyes and ears. He lay there like he was sleeping.

*Voices and steps*, Storm remembered. Now he understood Alice's choice of the date. *There's a vacuum down there. Deep space...* Not that loud, they can hear us. *You're overworked. It's your guilty conscience about O'Brian.*

"I'm sorry. I didn't mean to," he murmured before leaving O'Brian.

A few rows away he found the man he was looking for.


*Oliver Storm*
*Police officer*


stood there. Instead of a seat tub, there was a bed. Rather, a cot. It would have been enough for an infant.

So there he was, the real Commissioner Oliver Storm—lying on a special cushion. Despite the beard and the parts of his face covered by helmet and Rift, he recognized himself—the angular nose, slightly bent to the left, the lip area. His head was fixed with a face cradle to prevent him from rolling down. The severed neck was overgrown, scar tissue grew like a monstrous light-colored hair fuzz under the full beard. Protruding from the scar were red, blue, and beige plastic tubes, which were connected to a machine with valves and fittings. A stainless steel frame, cables, needles, sensors, flashing LEDs, medicine dispensers for neurotransmitters, and psychotropic drugs. The machine hummed softly, the tubes moved rhythmically as blood and lymph were pumped through the head of Oliver Storm, police officer.

There he lay, comfortably bedded, dreaming of a starship setting off for a new world. Dreamed in the dream that he had been beamed back into the here and now of reality and was watching himself dreaming.

Storm struggled with the impulse to touch the lost head that lay there. So who was touching whom? To be precise, there were three Oliver Storms, but all three were nothing more than neural patterns in the head of the One, the dreamer who lay there. He pulled his hand back. Was it a cosmic farce or a philosophical riddle? Either way, it did not seem right to do so. He turned away and followed the flashing directional arrow that showed him the way out of purgatory.

Again, Storm found himself in a no man's land. Three levels of nested illusions, deceptions, charades. The only real level was so unreal that his mind refused to recognize it as such. What did a head—a singular head in a secret military research facility, a head deprived of all sensory impressions—have to do with Oliver Storm? The head was not alive, you couldn't call that life, because it did not interact with its environment. It had no environment at all. It was an antenna for a reality created in a gigantic computer. The idea of being at home in a machine was oddly reassuring.

A strange atmosphere prevailed in the next few days between the initiates, as he now called his friends. They went about their daily routine. No one mentioned what they had seen in the Forensic Time Machine. They avoided the subject with casual obstinacy. Not even with Nordica did he talk about it. Should he say "I visited myself"? Or "I'm lying there on a lab bench... I'm nothing but a clean-cut head," and then they would laugh about it. "A good head," that was even better. Should he ask if she had entered the restricted zone like him, looking for what was left of her? Not even the harmless question of what she thought of all this came across his lips. It was impossible for him to talk about it, the reality was unspeakably far from anything imaginable. And whereof one cannot speak, thereof one must be silent.

| | |
|---|---|
| Distance to destination | 0.000 ly |
| Time to landing | 2 d |
| Distance from Earth | 4.243 ly |
| Speed | 0.000 c |
| Acceleration | 0.000 g |
| Slip correction | 334.50 d |

## Episode 19 Landing

#news flash 16.3.2224
Apophis II is expected to hit Earth in two months' time. The bunkers of the Mountain Fortress project have been completed. All moon bases will be

massively reinforced. We hope to maintain contact with Atlantis. Our best wishes accompany the colony.

© RG channel via moonbase Clarke II, sent 16.12.2219

Two more days until landing. The final orbit brought them within 2000 km of Atlantis. From up here, their new home did not look inviting. The hot dayside shone in the pink light of the central star like the oiled skin of a sun worshipper. The dark night side showed huge glaciers and icy seas in the infrared image. The habitable area along the twilight zone consisted of mountains, rolling hills, rugged cliffs, and waters—strangely shaped lakes of all sizes dotted the terminator belt of the planet, and above them mighty cloud formations were drifting. Here and there, brown-green patches lit up between the mountain ranges. Somewhere down there was Shangri-La, the Promised Land. The tiny moon circled the planet in a close orbit. It looked like an artificial satellite, attentively watching its host.

The crew was constantly busy with all kinds of things. Storm's team prepared for the landing. Everyone was eagerly awaiting the big day, which would actually last for several days, but at the same time, the enormous workload was eclipsing the upcoming event.

The landing preparations were routine. The scientific and technical modules were automatically landed one after the other. Then it was the turn of the accommodation containers. Finally, the actual cargo was unloaded—material, supplies, spare parts. Storm's troops supervised the process, so that they could intervene if necessary, as they had practiced for dozens of times. Everything went perfectly, almost too perfectly for the inspector, who had expected intervention from the programmers. But nothing went wrong; maybe it was because the Ministry had now taken over the simulation planning, or maybe it was because Ahlgrim was out of the game and could not thwart anything.

And then the time had come. First, crew 7 was to leave the ship to prepare the colony for the other crews and the passengers, who would gradually be brought out of cryosleep. Storm's crew was to collect air, water, and soil samples before the actual landing maneuver—for the first time in VR.1. He put together two teams: Tom and Cat formed the first, he and Nordica the second. They landed with the Mayflower and Discovery on the windswept plateau of Cape Cod. Storm paid attention to detail and in fact the landing seemed more real to him than the numerous simulations they had performed over the years, although he knew there was no difference. Reality was not only fragile but also capricious.

The rovers quickly took them down to the valley where they accomplished their tasks. Tom and Cat drove on to the lake where David had once sunk entire housing containers, while Storm and Nordica remained in the settlement, checking the modules and analyzing the air on the hill with the portable mass spectrometer. It was not without ulterior motives that he had distributed the tasks in this way. It would give Alice, who no longer had to hide in VR.2, the opportunity to talk to them.

They climbed up the hill. Storm took the samples on the summit. Nordica admired the landscape—bushes, rolling hillocks, the Grampians in the distance, the waterfall, the creek at the bottom of the valley. It looked exactly as it did in VR.2, and knowing this could not destroy the intense impression that this could be their new home if only they wanted it to be. Not a land of promise, but a fixed point that resisted the doubt of perception.

The Commissioner had made his decision.

Minister Rathenau, too.

"Hello," said Alice, who materialized behind a bush. Nordica was startled but recovered quickly. Storm formally introduced the two women to each other as if they were at a party. It was absurd to introduce a just appeared ectoplasm with the words "This is Alice Falkenberg, my former colleague," but they were all ectoplasmic genies in an absurd simulation, and yet everything was real again.

Alice got straight to the point. "What did you decide? Are you staying here?" she asked, as if they wanted to go for a little walk or not. Storm looked at Nordica and lowered his eyes. He hadn't dared to talk to her about it. What if she didn't want to stay? What if she accepted Alice's offer to leave and take up a Waldo existence in VR.0? Maybe VR.0 was a simulation too, it flashed through his mind. Through which head, it was hard to tell. He felt existential panic gripping him—the feeling of not being real, of not knowing if anything was binding the ego, the doubtful ego together.

Alice had noticed his anxiety. She paused and looked over the top of the hill, checking to see if the rover with Tom and Cat was there. Over there was the lake.

"EDDA intended to stop the game after arrival at the destination. Apart from the topography of Shangri-La, there is no data about the planet in the system, no scenarios, nothing. You would arrive with the rovers after ten kilometers at a border where nothing is behind—a gray wall."

"Just coordinates," Storm muttered knowingly.

"We have a plan. Our VR geeks will create beautiful things for you," Alice continued.

Storm looked at Nordica. Was there approval in her eyes, or was it just fatalism?

"I think this is a good place," he said. When it was said, he wondered how easy it had been. A soft smile played around Nordica's lips. She nodded approvingly as if it had always been clear.

"What about Jason and David? Have they seen the Waldos yet?"

"Sure. And maybe a little more," Storm murmured vaguely. Alice's reaction made it clear to him that she had suspected, maybe even hoped, that the four initiates would enter the military restricted area anyway, where they would find the remains of themselves.

"We lost Dana," Storm noted.

Alice nodded. "She just left in a landing capsule. When she ran out of oxygen, we retrieved her. She told us about an old film that inspired her to escape. She's fine. She's taking care of Zigmund's lab."

From a distance, they could hear the sound of the rover coming back from the lake. Alice approached them, grabbed their hands, and squeezed them.

"You must start the landing operation tomorrow, not two days later as planned. Crew 7 will be completely disembarked. Also, some passengers to freshen up the settlers. Molander is prepared. And tell your friends, if they want to leave the simulation, tell them to sign up for the last landing capsules! For the last ones, will you?"

"Ok?" said Storm, waiting for an explanation that didn't come.

"I'll visit you sometime," she said and became transparent.

Back on board, they reported to Molander that all the modules were working. The sample analyses were consistent with the results that the unmanned Starshot probes had sent to Earth a long time ago. The air was breathable, the water was germ-free and the soil morphologically, chemically and microbially resembled a chernozem similar to prairie soils. (Anyone else would have been surprised, but Molander the softie couldn't be surprised and Storm knew it was programmed that way.)

The next day the first passengers disembarked. Each landing capsule held three people. They were gradually released when the EXODUS was in a favorable position. Storm had told Jason Nygard and David Müller about Alice's story and transmitted her instructions. David had decided to get out of the simulation. He was too old to start a new life again, just pretending. The option of returning to reality put Jason in a philosophical crisis. He wished to and did not want to. They listened patiently to what he had to say, but his arguments went in a loop: he would like to stay here with his friends. On the other hand, he wanted to go back to the real world, there was no question, they had to understand that. But would life as Waldo be more real than this one? If not, then he might as well stay here with his friends, who are, however, only virtual friends. And so on...

The inspector wondered what Alice's instructions meant. Did they want to crash the last capsules in order to have an excuse for the disappearance of some crew members by a faked catastrophe? Or should they officially stay on the EXODUS, in orbit around Atlantis? And what was to happen to the virtual passengers who were still asleep? Were they all softies who would simply be switched off? A pretext for their disappearance could easily be found.

Then came the time to say farewell. He himself, Nordica and Jason, who had chosen the new world thanks to Aristotelian logic, set off for Atlantis.

"Good luck," whispered Storm to David as they embraced. "I will miss you."

"Maybe I'll visit you some day," David said confidently. Certainly, in this direction—from VR.0 to VR.1—it was possible. That made it easier to say goodbye.

The capsule with Henderson, Nygard, and Storm landed on Atlantis shortly after midnight. Proxima was ten degrees below the horizon. The orange-red band of dawn spread above the Grampians. The sky over the hills to the east was still dark, Perseus and Cassiopeia stood high. Maybe they should rename the constellation *Cassiopeia Nova*, Storm mused, because Sol shone brightly there like a nova.

It was a long walk to the colony, across the plateau to the south, then steadily downhill. Halfway down it suddenly became brighter. On the horizon, a northern light flickered like the reflection of a distant fire.

"We've got a flare," Molander reported minutes later. "How long till you reach the colony?"

"About thirty minutes," the inspector estimated. "Twenty, at least."

"Too long. The tempest will get you in fifteen, at the latest. I'll have someone pick you up."

Storm and Nygard discussed the technical issues. *Absurd*, thought Storm, shaking his head in confusion. *Why these games?* Nordica had watched him intensely. She understood immediately.

"It is not straightforward to intervene with sophisticated simulation software. Things like that are programmed, even stochastically."

Nygard looked up in surprise. "Did I miss something?"

"Not at all, my dear Watson," she teased him. "Just tell us if we can reach the colony in time."

They agreed to try it on their own. Their particle shields would be sufficient for a brief period in the open. Molander agreed, and so they moved on quickly in full protective gear.

When the northern lights flared up, the three of them shouldered the ring coils and activated the magnets. Green and blue veils flitted across the firmament. The settlement was near; in the semi-darkness, they recognized the directional antenna on the hill, some containers, and the huge particle protection shield of the settlement. Like a fairground wreath, it flaunted on the high mast. They began to run.

"We can do it," gasped Nygard. "We'll be there in ten minutes. That's maybe two hundred millisieverts. A cinch."

From afar, they could see the people in the shelter under the powerful magnet; they waved and shouted something to them. When a curtain of light from the Aurora swept over the settlement, the air above the six-meter-wide ring lit up like the signal of a lighthouse.

Less than two hundred meters from the settlement, Storm stopped his companions, catching his breath.

"We shouldn't run along," he admonished them, "even if it means a few fake millisieverts more for us."

Nordica in the middle, they approached the settlement, striding side by side at a proper distance to keep the force between the magnets small. The supra-coils transformed the proton storm of the stellar wind into three hoops of luminous plasma that floated above their heads. Like the Three Wise Men they moved into their new home, applauded by the settlers.

Orderly chaos reigned in Shangri-La. Everyone was running around in confusion, but everyone had their job and knew what to do. The modules for power supply, water treatment, communication, and the cold storage for the gene banks were activated, the large main antenna was tested, the particle shields were set up, the rovers and the two shuttles were stationed in hangars, the supplies and spare parts were sorted, and the dwelling containers were aligned and stabilized according to the plans.

Storm learned from EXODUS that the remaining landing capsules had to be checked, there had been a malfunction of the launchpad. One of them was to land the next day with Angelis, Müller, and the doctor Alex Jenner.

They will crash it, he realized. Somewhere on the night side would be best, because no one would ever find the wreck there. Two softies and the hardie David, who would lead a real life after death. A handful of pilgrims were still on board. Storm wondered why they were not yet dispatched.

During the night the inspector's smartphone shrilled. Drunk asleep he took off.

"Molander here. There is news. Our radar shows a cloud of asteroids approaching rapidly. It is possible that some of them may be heading for Atlantis."

"What can we do?"

The captain was silent for seconds. "Nothing. Stay on alert."

Storm gave Nordica the basics and informed the responsible NCOs. The Mayflower and Discovery were made ready for take-off. Then he woke up Nygard, who constantly shook his head after he was responsive.

"Now of all times? There is no asteroid belt in this system. Something is fishy about it. What speed? Hyperbolic or orbital?"

The inspector had no idea. He contacted the EXODUS. "Do you know anything more up there? Nygard wants to know the speed."

"Yes, the cloud is heading straight for us. Relative speed 0.15 c."

Nygard had overheard.

"0.15 c??? Holy shit! What's that? Where's the radiant?" It didn't take Molander long to answer. "Alpha 2/29/43, Delta 62 degrees 40 minutes 46 seconds."

Nygard worked on his mablet. "That looks damn familiar," he muttered.

"We've already checked it out," Kvalheim replied into the silence, which was only disturbed by Nygard's typing. "They come from Earth."

Nygard froze in the movement. By his gaze, Storm could tell he was thinking. Suddenly he raised his eyebrows. "Wait! This is—this can't be..."

Again he hammered frantically on his mablet. After a few minutes, he had a result.

"It's the JIAN TOU." He put the device aside. "Or what was left of her after we destroyed her. It all fits—speed, direction, time of arrival. The fragments flew on toward Atlantis after the explosion, at 0.15 c. Now they have reached their destination."

"When is it due?"

"Two hours."

After a second of shock, Storm picked himself up. "Okay, what can we do? What are the chances of getting hit?"

"Not very likely," Molander reassured him. "The radar only shows fragments larger than a meter, roughly. An estimated thousand of them, spread over a scattering cone several tens of thousands of kilometers in diameter. The probability that such a fragment will go down within 100 km of Shangri-La is one in ten thousand."

"And the smaller ones?"

"Hard to say. Maybe one in a hundred. But they are harmless, at least as far as ejection is concerned."

Storm pulled his shoulders up questioningly. Nygard helped.

"This means that only the very big lumps of debris during impact will throw enough material into the atmosphere to trigger climate change."

The inspector laughed. This time it was Nygard who did not understand.

"Climate change? Here? Very funny, don't you think?" Storm asked sarcastically.

Those two hours were the longest of their lives. At the bottom of his heart, Storm knew that nothing dangerous would happen; the Ministry of Peace had decided to maintain the colony. Still, he couldn't help feeling forlorn, so deeply had VR.1 infiltrated his mind. He threw a veiled glance at Jason who pretended to be too busy with his equations to look back. *All right*, Storm thought. *You are infiltrated like me. And so is Nordica.*

The Commissioner didn't want to inform anybody on the crew except the NCOs on duty, but Nordica convinced him that this was not a good idea. Even if nothing happened to them, the non-information would be interpreted as dishonest. So they woke everyone up. There was no arena like on the EXODUS or a meeting room, so the briefing took place on the hill. Nygard explained to the assembled crew that fragments of the JIAN TOU were approaching at high speed, possibly hitting Atlantis. His probability statements could not really reassure the crew. Therefore, Storm took the floor. He explained that the worst that could happen was an earthquake far away. If there was a threat of a tsunami on the river, they would reach the plateau with the rovers. There, the Mayflower and Discovery were ready to evacuate to the mothership if necessary. The Endeavor was parked in orbit.

They waited in silence, because there was nothing else to say. When the opportunity arose, Storm murmured toward Nygard: "Why the hell don't they remove this debris from the game?"

"We'll soon know," Nygard replied casually, without interrupting his skywatching.

And then the first falling stars came. A rain of tracers covered the night-dark sky, rising over the hills and disappearing to the west, where Proxima, still below the horizon, weaved a green-gold band of dawn.

They waited for the impact. A roar, a thunder, a trembling of the planet perhaps, who knew what it would feel like. But they were spared. The meteorite fall subsided, a gentle morning breeze rose, the river gurgling from afar. Everything was peaceful.

Storm looked up to the mothership that passed majestically above their heads. A brilliant arrow, pushing the shimmering spider web of her antenna like a bridal veil in the wind. The Endeavor accompanied her as a faint dot.

Then he spotted another arrow, staggering and distorted, rapidly approaching the EXODUS. The tattered antenna was barely visible—the wreck of the JIAN TOU on a collision course. A lightning bolt plunged the landscape into a ghostly blue. When the nightmare faded, Storm searched in vain for the EXODUS in the sky. The blinking fragments of the proud ship drifted apart like the sparks of a cosmic firework. *So they end the story*, it flashed through Storm's mind. *The interface to Earth destroyed and a perfect exit point for David and the pilgrims who were considered unfaithful.*

And thus Oliver Storm, Nordica Henderson, and Jason Nygard from the legendary Crew 7 were the last to leave the starship before the catastrophe. In the myths of later generations, they would be called the Magi. But that is a different story and shall be narrated some other time.

# 2

# The EXODUS Incident: A Failure Analysis

*Technical Report to the ESA inquiry panel*
*Classified document—do not distribute*

*RealGames GmbH—Internal revision*

**Executive Summary**
This report describes the results of an investigation into programming errors of the EXODUS virtual reality. To this aim, the physical basis for the codes EXO VR.1 and VR.2 (internal abbreviations for the EXODUS virtual reality program codes at levels 1 and 2) was checked. It was found that several task groups involved in the programming used oversimplified theoretical models, neglected physical details, or ignored the internal communication protocol. In view of the fact that the European Ministry of Peace commanded the continuation of EXO VR.1 on Atlantis, it is recommended that each task group be strengthened with experts, and a supervising group be installed that enforces the communication between the task groups.

© The Author(s), under exclusive license to Springer Nature Switzerland AG 2021
P. Schattschneider, *The EXODUS Incident*, Science and Fiction,
https://doi.org/10.1007/978-3-030-70019-5_2

**Contents**

Introduction......................................................................................................... 159
The Teams............................................................................................................. 159
    Ramjet............................................................................................................ 159
    Space Travel.................................................................................................... 162
    Construction................................................................................................... 163
    Weaponry....................................................................................................... 168
    Astrophysics.................................................................................................... 169
    Chemistry....................................................................................................... 177
    Atlantis........................................................................................................... 177
Conclusions.......................................................................................................... 181
Appendix: Technical details.................................................................................. 182

# Introduction

RealGames GmbH is a contractor of EUROFORCE. The contract covers data processing and programming of the environments EXO VR.1 and EXO VR.2 on board the virtual starships EXODUS I and II and on the virtual planet ATLANTIS. RealGames GmbH takes the opportunity to declare that at no instant RealGames GmbH was aware of the malversations and atrocities committed by agents or business partners of EUROFORCE. We address our sincerest commiserations to the victims of the EXODUS incident.

In the lawsuit filed by the Ministry of Peace against members of EUROFORCE, the defendant claimed that errors in the EXO VR.1 and VR.2 codes led to the disastrous outcome of the experiment. The records show indeed that the physicist on board detected phenomena that were inconsistent with physical laws such as the wrong relativistic time slip or the missing plasma aureole at the fuel entrance. His observations supported suspicions that the crew lived in a virtual environment, and led to the disclosure of the project's hidden intention.

For this report, it is important to understand that RealGames GmbH never pretended to simulate a realistic future. The task was to construct a "what-if" scenario that should be consistent with the laws of physics. A number of physical parameters, in particular for the not yet existing fusion thruster or the Atlantis environment had to be assumed. Guidelines and specifications for EXO VR.1 and EXO VR.2 were stipulated by the customer. The software was finally delivered to THZ S.A.R.L. It was not in RealGames GmbH's competence to check what happened there and afterward.

The code was written by seven independent teams. The main tasks of each team are described below in order to show how misunderstanding, communication problems, or too coarse physical models led to the well-known problems and the unexpected discovery of virtual reality by some crew members.

# The Teams

## Ramjet

The main task of the team was the design of the fusion thruster, which converts interstellar hydrogen into helium. The group used the original concept of magnetic scooping proposed by Robert W. Bussard in 1960. Strong electric and magnetic fields at the bow of the spacecraft would act like a funnel compressing the interstellar gas of ionized hydrogen into the axial pipeline of the ship.

Despite numerous publications, no electro- or magnetostatic arrangement is known so far that would allow the collection of interstellar protons over an area of many square kilometers and their subsequent compression. We, therefore, assumed—like other authors did—that a suitable field configuration can be generated, and delegated this task to *Construction*. In order to prevent a possible query in the ship's archives, we declared the alleged description of the magnetic funnel a military secret.

The compressed hydrogen would serve as fuel for the fusion reactor. The construction is similar to a ramjet used in aeronautics: The forward motion of the ship relative to the interstellar medium is sufficient for the compression. The faster the ship is, the higher the flow of the supplied fuel and the stronger the thrust.

The ionized hydrogen is fused to helium in the reactor at a temperature of several hundred million degrees. In this process, as in commercial fusion reactors, the potential energy of the atomic nuclei is converted into kinetic energy, which results in a higher ejection speed of the fusion product and thus generates thrust. Commercial reactors use heavy hydrogen (deuterium D) as the starting product, which is only present in traces in the interstellar medium. Therefore, a deuterium nucleus must first be produced from two protons: p + p → D. This process is 24 orders of magnitude slower than the fusion of two deuterium nuclei to form helium and is therefore unsuitable for generating effective thrust. Therefore, the Bethe–Weizsäcker or CNO cycle is used in the simulation, a catalytic process that uses nuclei of carbon, nitrogen, and oxygen as intermediaries in order to produce a helium nucleus out of four protons. The CNO cycle is the dominant process in hot main sequence stars and to a lesser extent in the Sun. It is 18 orders of magnitude faster than the p + p → D reaction. The amount of the catalysts in the fusion chamber amounts to about 1% of the ship's mass.

The fusion drive has been designed to generate a relatively low peak thrust of 1.7 million Newton, which is one-twentieth of the thrust of the first stage of the ancient Saturn V rocket. The volume of a sufficiently powerful reactor corresponds to a sphere of about 11 m radius, making up a toroidal chamber of slightly higher volume. It "burns" about 800 g of hydrogen per second. This guarantees a maximum acceleration of 1.18 $m/s^2$, which corresponds to a perceived 0.12 Earth gravity on board.

The *Astrophysics* team provided the density of the interstellar gas outside the heliosphere, which is very low with 0.5–1 million hydrogen atoms per cubic meter. This results in a completely unrealistic sweep radius of over 2000 km for the intake funnel. Since this calculation is easy to perform even with little knowledge of physics, we decided to assume a much higher density of one hundred million hydrogen atoms per cubic meter, resulting in a more realistic sweep radius of 200 km. Unfortunately, the technical literature is unambiguous on this question—the solar system is located in an interstellar cloud of very low density. To ensure the illusion, we included faked results from the more recent Starshot missions in the imaginary library, that collection of fictitious publications from the twenty-second century. It is claimed that these have documented a density of the local interstellar cloud extending to Alpha Centauri that is more than a hundred times higher than reported in the grossly deficient publications from the twenty-first century, based on old data from the Pioneer and Voyager space crafts.

Interstellar gas consists of 95% of hydrogen, two-thirds of which are ionized, and one-third are neutral $H_2$ molecules. An electron stripper, following the well-established principle used in ion thrusters, here in the form of a positively charged forest of carbon nanotubes ionizes the neutral components, which may then also be harvested with magnetic forces. The ramjet would reach 93% of light speed within 13 years. However, the maximum velocity on the journey to Proxima Centauri is 56% of light speed, attained after 7 years. During the next seven years, the ship slows down gradually using the magnetic funnel as a braking sail until it reaches Atlantis. The fuel supply and the energy gain from fusion as well as acceleration and speed shown in Fig. 2.1 were calculated relativistically. No error was found in the derivation.

It is perhaps surprising that the onboard acceleration drops after about two years. This is a relativistic effect: the fuel protons' relativistic mass increases with the ship's velocity, according to Einstein's equation. Therefore, the energy gain from fusion results in a lower gain in linear momentum per proton.

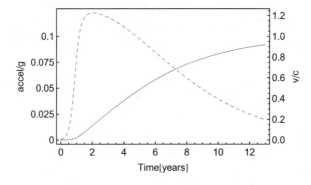

**Fig. 2.1** Continuous acceleration (dashed) and speed in units of light speed c (full line) as a function of proper ship time for a ship mass of 1500 tons, a sweep radius of 200 km, 15% efficiency of energy conversion into thrust, and an initial velocity of $10^{-5}$ c

Therefore, the ship's speed approaches light speed without ever reaching it. Seen from earth, the reason for never reaching the speed limit is the ever-increasing mass of the ship.

***Failure Analysis*** An inquiry regarding the design of the proton collector and the superconducting coils was sent to *Construction*, which in our opinion was responsible for the technical implementation. It turned out that *Construction* forwarded the request to *Atlantis* who in turn forwarded it to *Astrophysics* where it got stuck. This led to the missing plasma discharge phenomena on the entrance to the pipeline.

## Space Travel

The team's main task was the design and control of the journey to Atlantis.

### Basic Equations

Knowing the thrust F of the engine and the mass M of an object, the acceleration is $a = F/M$, according to Newton's second law. When $a$ is known for any instant of the journey, it is easy to calculate at which distance $x$ from earth the ship is at any time $t$ by solving $a = \dfrac{d^2x}{dt^2}$ for $x=x(t)$. This function is called the *world line* of an object in the theory of relativity. The tricky part comes from relativistic effects. At high cruising speed, clocks on the ship run slower seen from earth, and the Lorentz contraction changes distances. That makes $x$ and $t$ observed in earth-based coordinates different from ship-based coordinates. One solution to this problem is to calculate both $x$ and $t$ as functions of the proper ship time $\tau$. This yields the world line and also the difference between earth time and ship time—the *slip correction* displayed on the Belt monitor.

***Failure Analysis*** Originally, the relativistic *slip correction* was calculated with an accuracy of seconds. The request of a GHz frequency standard from the EXODUS depot alarmed the steering committee, and *Space Travel* was charged to rewrite the respective source code of EXO VR.1 with microsecond accuracy, then implemented as Patch 27.2. When it was realized that an atomic clock was positioned on level 4 of the rotating reference frame where an additional time dilation effect from general relativity not considered in Patch 27.2 occurs, the steering committee decided not to change the code because the effect was considered to be too small to be measured. As we know now, we underestimated the ability of the physicist on board.

## Destruction of JIAN TOU

Relying on low thrust technology, the Chinese starship JIAN TOU was scheduled for a flight time of 31 years. It was launched 13.5 years before EXODUS II, which passed JIAN TOU during the deceleration phase where the relative velocity of one to the other was 15.6% of light speed.

Figure 2.2 shows the world lines of the two starships. Figure 2.2b zooms into the time after the destruction of the JIAN TOU. Its debris arrives at Atlantis 54 days after the arrival of EXODUS II. The thrust, encounter, and destruction of the Chinese ship 3.7 light-years away from earth were carefully tuned so as to guarantee the precise timing of the debris impact on the orbiting EXODUS II. No failures were detected in the protocols. Note: the original plot envisaged the termination of the experiment on day 54 by destruction of the ship with the entire crew on board. When the Ministry of Peace took over, evacuation was ordered two days earlier.

## Construction

The team's main task was the design of the ship, the antenna (called spider web), the GCR shielding (against galactic cosmic rays), protection against heat, and radioactivity of the fusion drive.

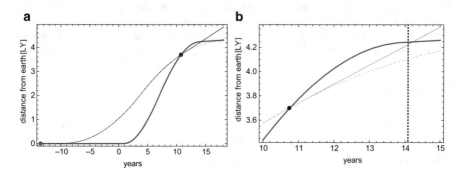

**Fig. 2.2** (a) World lines of EXODUS II (bold blue line) and JIAN TOU (thin line). Time scale measured from the start of EXODUS II on 1.1.2210. The red and black dots mark the start and the destruction of JIAN TOU which otherwise would have followed its approach to Atlantis (dashed). (b) Zoom into the last four years. The dotted vertical line marks the arrival of EXODUS II in orbit of Atlantis. 54 days later it is destroyed by the impact of the debris (thin line)

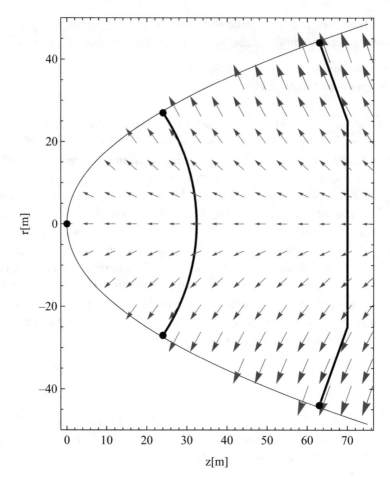

**Fig. 2.3** The Igloo and Belt section of EXODUS. Dots mark nominal gravities of 0.1, 0.2, and 0.3 g, assuming 0.12 g linear acceleration in positive z-direction (to the right). The bold lines mark the separations between Igloo, Belt and living quarters. The pseudo-gravity is shown as a vector field. It is perpendicular to the surface of the paraboloid

## Kinematics

The ship spins over the long axis once every 25 seconds. The centrifugal force creates a pseudo-gravity that increases with increasing distance from the rotation axis. The Belt and the medical unit are the areas with the lowest pseudo-gravity. Superimposed on that force is the acceleration of the ship which makes the direction of the sum of the two forces dependent on the distance from the axis. The habitat is shaped as a paraboloid. This form of construction guarantees that still-standing persons always stand on a horizontal surface as sketched in Fig. 2.3. The dots mark nominal gravities of 0.1, 0.2,

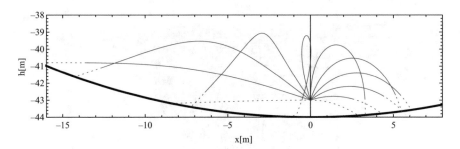

**Fig. 2.4** Jump or sprint width on the big promenade as a function of direction for a medium trained person. Ordinate: Distance from the rotation axis. Max. jump height is ca. 4 m, widest clockwise step ca. 14 m

and 0.3 earth gravities under the assumption of a linear acceleration of 0.12 g. The *Igloo* with the medical unit and the tanks is located in the leftmost section. The Belt with the limiting promenades is located between 0.2 g and 0.3 g gravity, chosen according to experiences after the Mars missions that showed that Mars' gravity is sufficient to avoid bone and muscle degeneration. The diameter of the big promenade where 0.3 g prevails is 88 m.

Running on the promenades is easy when going clockwise (against the habitat's rotation). Figure 2.4 shows jumps of a medium trained person (who succeeds a vertical jump of about 0.3 m on earth)—at the big promenade, in different directions from horizontal clockwise (to the left) to horizontal anti-clockwise (to the right) in steps of 18 degrees. Drawn are trajectories of the center of gravity of a sprinter/jumper for one step/jump, assumed to be located 1 m from the floor. The dashed lines continue for a hypothetical pointlike mass. Clockwise, a jump of about 14 m is feasible, whereas anticlockwise about 5 m is the maximum. This applies for the big promenade with an artificial gravity of 0.3 g. This strange behavior is due to Coriolis forces occurring in rotating coordinate systems. The force on the jogger depends on speed and direction. This has interesting consequences for the popular game on board: Throwing a tennis ball such that it makes a full circle on the big promenade going back to the pitcher like a boomerang. Numerical calculation yields a necessary initial speed of 40 km/h. (The maximum achievable baseball pitching speed is about 100 km/h.) The linear acceleration of the ship obliges the pitcher to throw the ball on a skew trajectory toward the bow (Fig. 2.5).

## Atmosphere

Originally, an earthlike atmosphere of 890 Hectopascal was simulated, corresponding to a standard pressure at 1000 m over sea level on Earth. Pressure

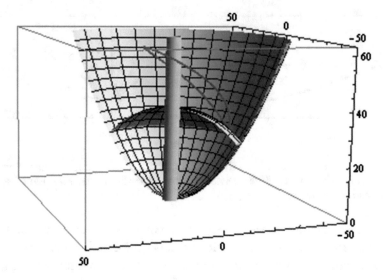

**Fig. 2.5**   Red line: A boomerang pitch on the small promenade. We see a cut through the hull, the Igloo, and the axial pipeline. Scale in meters

increases from the axis outward because of the pseudo-gravity, as on earth. The barometric height equation was implemented. But this equation does not hold for artificial gravity in a rotating environment.

***Failure Analysis***  When a request came from the ship for a barometer and an anemometer, the steering committee was alarmed and ordered a reprogrammed more realistic atmosphere for a rotating environment, implemented in Patch 27.11. Up to an axial distance of 330 m, the pressure increase is below that of an earthlike atmosphere: up to the radius of the big promenade, the pressure increases by less than one Hectopascal whereas on Earth it increases by 5 Hectopacal. Even with a very precise barometer, it would have been difficult to distinguish the two cases.

At this occasion, it was realized that winds could pose a problem for the maintenance of the illusion. In the original software, the sensation of winds was based on a random generator whereas in a rotating environment, according to Coriolis forces, circulating atmospheric cells occur similar to rotating depression systems on Earth. After a request of the steering committee, an estimate showed that the Coriolis forces on the airflow in the Belt cannot be neglected. A fast real-time finite elements code was then used to simulate turbulence. Results were implemented as Patch 27.10.

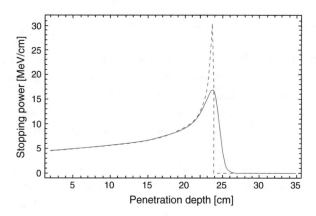

**Fig. 2.6** Stopping power of water for 194 MeV protons. At the Bragg peak (24 cm penetration depth), the particles come to an abrupt halt (dashed line). Stochastic broadening causes a tail extending to 26 cm (full line)

## GCR Shielding

The interstellar gas impinges on the front of the ship with speeds as high as 56% of the speed of light. The kinetic energy of hydrogen atoms, which provide the largest part of the gas, amounts to 194 million electron volts. By construction, the hydrogen ions are harvested by the magnetic field and compressed into the pipeline where they cannot harm the crew. Nevertheless, a small fraction escapes the harvesting. Such particles penetrate the thin metal/carbon composite hull of the starship. In order to get rid of them, the water tanks are located in front of the habitable section, building a 25-cm-thick protective sheet. The high-energy protons penetrate into water. The kinetic energy which they lose per unit path—the stopping power—is usually given in MeV/cm. Starting with 194 MeV, protons penetrate 24 cm before they come to a halt. The deceleration peak, also known as the Bragg peak, occurs in the last 2–3 cm as shown in Fig. 2.6.

The rare particles with higher kinetic energy can penetrate the water tank and are further slowed down in dense matter such as walls. Only a minuscule part of them reaches the crew where they are supposed to trigger GCR events.

*Failure Analysis* The water sheet of 25 cm was calculated without stochastic Gauss broadening (dashed line instead of full line). This mistake was not critical but it could have been noticed because such calculations are standard, e.g., for proton therapy of cancers.

## Faraday Cup

The experiment with the Faraday cup was a typical falsification experiment: Since it prevents electromagnetic waves to pass into the brain, virtual reality should not work. But it did. Consequently, the assumption that the helmet created the virtual environment was wrong. Curiously, a subtle detail escaped the experimenter: a Faraday cup blocks waves with wavelength greater or equal to the loopholes of a metallic grid. That said, Terahertz waves could have passed the Faraday cup—they are in the submillimeter wavelength range whereas the hand-made cup had loopholes of a few centimeters. At best, the waves would have been distorted by the metal wires, causing erroneous transcranial stimulation. In spite of a flawed argument, the Commissioner's group came to the correct conclusion that they lived in VR.1.

*Failure Analysis* A closer monitoring of the events on board would have prevented the disclosure of VR.1, by a feigned malfunction of EXO VR.2.

## Ramjet Scooping

*Failure Analysis* The request of *Ramjet* for design documents of the electromagnetic scooping funnel for interstellar hydrogen hit us unprepared because hydrogen compression was intended as part of the propulsion system and not of construction. Nevertheless, we sent a request to *Atlantis*, as this team had experience with superconducting coils for proton deflection. Unfortunately, our inquiry was never answered.

## Weaponry

The team's main task was the design of defense weapons against interstellar debris. Laser cannons were implemented for objects smaller than 10 cm, and Al metal spheres for larger ones. Up to 20 cm, debris would be vaporized immediately by the metal spheres, bigger objects would be disintegrated into smaller chunks that could be destroyed with the lasers. The steering committee requested verification of the lethal attack on JIAN TOU with simple metal spheres. The trajectories of the bullets were calculated on the assumption of an ejection speed of 3 km/s, the performance of railgun launchers on the Moon.

The kinetic energy of a spherical Al bullet of 200 mm diameter at an impact speed of 0.15 c—the relative speed between the two ships at the encounter—is almost 200 times the energy of the historical Hiroshima bomb. The projectile would penetrate the hull of the target and punch a hole about 1 m in diameter, several others on the opposite wall, cause cracks, destabilize the structure, and lead to an immediate drop in pressure. On hitting the reactor shield, it would disintegrate, releasing at least 50% of its kinetic energy and igniting local fusion processes which would inevitably destroy the reactor.

*Failure Analysis* No programming error could be detected, although some members of the panel criticized that launching a bullet when the target is at a distance of 200 million km, the chance of hitting it at a distance of 14,000 km is as probable as winning the lottery.

## Astrophysics

The team's main task was the physics of the interstellar medium, the star field, and the ship's exterior.

### Star Field

Seen from destination Atlantis, the Sun is a star of magnitude −0.05 located in the constellation of Cassiopeia, in the lower right of Fig. 2.7. Note the slightly distorted stellar constellations.

### Geometric Aberration

At maximum speed (0.56 c) the star field is compressed toward the flight direction because light rays seem to reach the observer more from the bow, similar to rain hitting a fast-driving car. That creates a denser and brighter star field. In the rear view, the opposite happens. Stars are fainter and appear under larger angles from the rear center. Figure 2.8 shows the visual distortion of a (hypothetical) uniform distribution of 100 stars on the sky along a complete circle around the starship. The geometric aberration makes each star appear under a smaller angle with respect to the flight direction. The visual appearance is shown on the inner circle, creating a denser star field at the bow and a less dense distribution at the stern. A comparison of Figs. 2.9 with 2.10 shows this stunning effect.

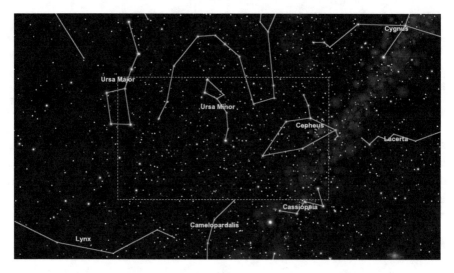

**Fig. 2.7** The Sun seen from destination Atlantis is the bright star of magnitude –0.05 in the Cassiopeia constellation. Polaris is at the center of the image. The field of view is 110 degrees × 70 degrees. The yellow frame indicates the effect of geometric aberration: at the maximum speed of 0.56 c, the star field inside the frame would appear blown-up to the full field of view. In Figs. 2.7, 2.9, and 2.10 the lower limit of visible stars is set to magnitude 6.5. Star field created with Celestia software

***Failure Analysis*** In the hearing, members of the committee were surprised by the dense star field in Fig. 2.10 and expressed doubts concerning the programming. Internal revision confirmed that this is an effect of aberration at relativistic speed. The only caveat is the fish-eye-like distortion of the wide-angle camera toward the outer parts of Figs. 2.7, 2.9 and 2.10.

## Diurnal Cycle of Proxima

The simulation of the diurnal cycle on the planet was based on data transmitted from the Starshot spacecraft. The planet has a tidally bound rotation, that means the duration of one day equals the duration of one year. In a circular orbit with the rotation axis perpendicular to the orbit, the central star would forever remain in the same position in the sky for an observer on Atlantis, such as the earth seen from the moon. Here, we have two subtleties: (1) The rotation axis of the planet is inclined by 15 degrees to the orbit. (2) The orbit is elliptic. Figure 2.11 shows the situation during one year (or one day, for that).

An observer on the equator (white spot in the hot day zone) would always see Proxima in the zenith. But according to Kepler's second law, in an elliptic

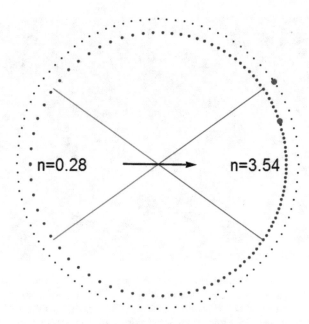

**Fig. 2.8** Geometric aberration, assuming an isotropic star field around the EXODUS. At the maximum speed of 0.56 c, each of the 100 stars uniformly distributed on the circle (tiny dots) appears shifted to the bow (big dots). The two red dots are an example. The area density in flight direction is 3.54 times higher than for an observer at rest. Against flight direction, looking back to Earth, the area density is reduced to 28% of normal. The blue and red cones indicate angles of ±35 degrees, the vertical extension of the bow and stern screens

**Fig. 2.9** Alpha Centauri (bright star in the center, magnitude −1.8) at mid-distance for an observer at rest relative to the Sun. Proxima is indicated by the red star symbol just below Alpha. At magnitude 9.6 it is not visible to the naked eye. Star field created with Celestia software

**Fig. 2.10** Same at maximum speed (56% of light speed). The geometric aberration compresses the star field in the viewing direction. The Doppler shift increases the apparent temperature of the stars and their brightness. Alpha Centauri (center star) appears at magnitude −4.5 whereas Proxima shines at magnitude 7. Star field created with Celestia software

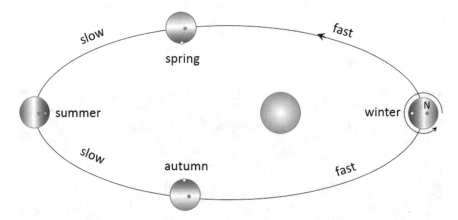

**Fig. 2.11** One year on Atlantis. The nonuniform orbital speed creates a periodic east-west deviation of the apparent position of the central star from zenith on the observer's (white spot) sky. The tilted rotation axis of Atlantis (see the decentered north pole in blue) causes a periodic south-north deviation. One stellar year, identical to one day is 269 hours. The orbital eccentricity is 0.11

orbit, the planet revolves faster when closer to the central star, and slower when farther away. Assuming that the rightmost position in the ellipse corresponds to January, the planet travels a longer distance than on a circular orbit during the first 3 months, and the next 3 months a shorter one. So in spring, Proxima appears east of the zenith. Until summer, the slower motion

**Fig. 2.12** The Grampian mountains, looking west from the landing site Cape Cod, 45 hours before sunset. During one day (135 hours) Proxima tracks the orange line, spanning an angle of 30 degrees horizontally. The yellow line is the track of the Sun if Earth would be tidally locked. Proxima appears three times bigger on the sky than the Sun would. Image: "Atlantis Landscape II" by Lucas Giesinger. [https://www.flickr.com/people/192057596@N04/], licensed under CC BY 2.0/A derivative from "The White Pocket" by John Fowler [https://www.flickr.com/photos/snowpeak/] and from "Solar Flare" by NASA, both licensed under CC BY 2.0

compensates the missing rotation angle, the star is again zenithal. Until autumn, Proxima will have moved to the west on the observer's sky, in order to catch up one full rotation in next wintertime. The second effect concerns the 15 degrees inclination of the rotation axis: The blue spot marks the north pole. It is decentered in top view because the axis is tilted. That induces a periodic north-south movement of Proxima on the observer's sky, exactly as on Earth: the winter sun appears more southerly than the summer sun, with a considerable difference of 46 degrees S-N between the two extreme positions. For Atlantis, the difference is 30 degrees, superimposed with the east-west oscillation of almost 30 degrees. On a tidally locked Earth, the east-west movement would amount to 8 degrees.

Figure 2.12 shows the landscape at the landing site, Cape Cod, with Proxima on the western horizon, and a sketch of the track during daytime (135 hours). For comparison, the track of the Sun during daytime (6 months) is shown in yellow, if Earth would be tidally locked.

## Doppler Effect

Similar to changes in the sound frequency of a moving ambulance, the Doppler effect changes the frequency of starlight seen from the ship. Stars change their color to blue when viewed in flight direction, and to the red in rear view. Inserting the relativistic frequency change into the equation for a star's radiation spectrum (to good approximation a black body spectrum) it is realized that the star seems to change its surface temperature and its brightness as a function of speed and viewing angle. The visual appearance of stars was derived from the black body spectrum at a given star temperature, defined by the red line (*Planckian locus*) in the CIE colorspace of Fig. 2.13. The color

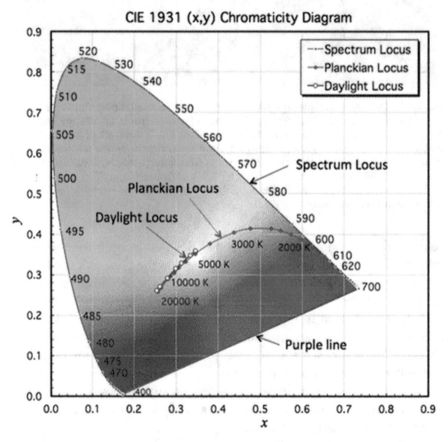

**Fig. 2.13** Color space. Red line: Color values of Black Body radiation (*Planckian Locus*). Rainbow colors are located along the outer border of the color space (*Spectrum Locus*). The popular idea of a *stellar rainbow* often described in relativistic space travel stories is erroneously based on the *Spectrum Locus* instead of the *Planckian Locus*. Reprinted by permission from Springer, Solid State Lighting Reliability Part 2 by van Driel W., Fan X., Zhang G. (eds) (2018)

variation along the Planckian locus is rather moderate and does not resemble the well-known rainbow colors. Stars with a surface temperature of 7000 Kelvin appear even white.

At maximum ship speed a sunlike star of spectral class G would appear as spectral class A in flight direction, whitish as Vega in the Lyra constellation (surface temperature 9600 degrees Kelvin), and to the stern as class M as Antares, the orange-reddish main star in the Scorpius constellation (surface temperature 3500 degrees Kelvin). The destination star, Proxima Centauri (surface temperature 3000 degrees Kelvin) appears at 5600 Kelvin, yellow as our Sun.

*Failure Analysis* The physicist observed the spectrum of the Sun with the prism spectrometer in the small telescope. Unfortunately, the Fraunhofer absorption lines were not programmed at that time. The problem was solved by rapidly programming a dysfunctional glass prism. During the repair time of the spectrometer, an emergency team succeeded in reprogramming the spectra of the Sun (Fig. 2.14), Proxima and Alpha Centauri with Doppler-shifted absorption lines. At 56% of light speed, the Doppler-shifted Fe T line goes to 562 nm. Thereby, the physicist correctly calculated the ship's speed. With Patch 28.1, the Doppler-shifted spectra of all stars down to magnitude 7 were implemented, including emission and absorption lines.

## Terrell Rotation

The physicist discusses the Lorentz contraction on the occasion of a dangerous close encounter of a sphere-shaped container from the destroyed Chinese

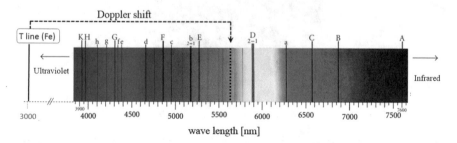

**Fig. 2.14** Optical spectrum of the Sun with Fraunhofer absorption lines. At maximum speed of the EXODUS, the Fe T line in the UV range is Doppler shifted to 562 nm (dashed line in the spectrum), close to the Na D line. The image is a derivative from "Fraunhofer-lines-in-the-solar-spectrum" by Shoude Chang, Yu Kui and Liu Jiaren [https://www.researchgate.net/profile/Shoude_Chang]/CC BY 3.0

ship. The Lorentz effect compresses a sphere into an ellipsoid with a short axis along flight direction. This was correctly implemented in the code. At a relative speed of 15.6% of light speed, the contraction amounts to 1.4%, too small to be seen with the naked eye. Interestingly, as the physicist points out, on photos or videos with near-instantaneous takes, the Lorentz contraction is not visible. Instead, fast-moving objects appear rotated. The effect comes from the well-known fact that the deeper one looks into space, the more one looks into the past. That means that parts of the object farther away from the observer (symbolized by the red dot in Fig. 2.15) are seen at an earlier time than parts closer to the observer (white dot). This elongates the visual appearance of fast-passing objects, compensating the Lorentz contraction. Visually, a fast-moving sphere appears always as a sphere to the distant observer. This effect is known as the Terrell effect or Terrell rotation.

**Failure Analysis** The Terrell rotation was not implemented, only the Lorentz contraction. The bug was only realized after the observation of the passing hydrogen tank. The steering committee eventually decided to keep patch 28.1 because the chance that the bug would be detected was considered negligible. The team points out that even Einstein believed that the Lorentz contraction can be observed visually.

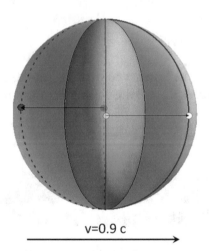

v=0.9 c

**Fig. 2.15** Lorentz-contracted sphere, passing from left to right at 90% of light speed. White and red dots on the centerline symbolize light sources at the near and the far end of the sphere. The red dot is seen at an earlier time than the white dot. To the distant observer, the central meridians appear rotated: The frontal one visible in blue, the rear one invisible dashed in red

**Interstellar Gas**

The most optimistic estimate for the density of interstellar gas—predominantly ionized hydrogen—is one million atoms per cube meter outside of the heliosphere, give or take 10%. The lower value of 0.8 million atoms inside the limits of the solar system is compensated by using supporting boosters and hydrogen fuel tanks at the beginning of the journey.

*Failure Analysis*  When *Atlantis* sent an inquiry about the density and composition of the interstellar gas, nothing was done because the data had already been delivered, albeit in another context to *Ramjet*. *Astrophysics* is not responsible for the fake density data in the fictitious library.

## Chemistry

The team's task was the support to other teams where chemistry plays a role: Composition of the atmosphere on board and on the target planet, reaction cycles, waste treatment, catalysis, biochemistry.

*Failure Analysis*  Radiocarbon, the radioactive isotope $^{14}C$ (and a proton) is formed when a nitrogen atom $^{14}N$ captures a slow neutron. Such neutrons are constantly produced in the Earth's atmosphere by cosmic rays. On board EXODUS, the production rate of neutrons is almost zero because the GCRs are so rare. So, there is practically no radiocarbon on board. Therefore, the organic material on board contained only the naturally occurring isotope $^{12}C$. Radioactive decay processes were not programmed in EXO VR.1 and EXO VR.2. It was not anticipated that someone could have the outlandish idea to apply the $^{14}C$ method to the vegetables the color, consistency, and taste of which we programmed so meticulously.

## Atlantis

The team's main tasks were landscape shaping, atmospheric phenomena, radiation shielding, and the design of life forms on Atlantis. The planet orbits Proxima in 11.2 days, at an average distance of 7.2 million km. It is tidally locked, turning always the same side to the mother star.

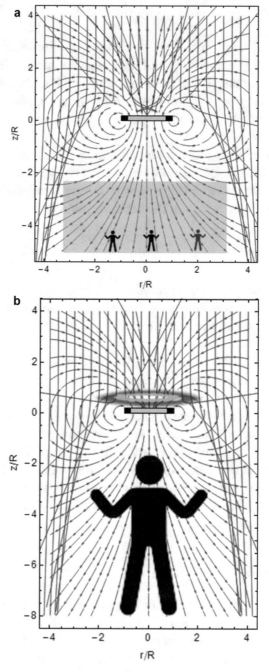

**Fig. 2.16** Superconducting particle shields. Blue: Magnetic field lines from a horizontal superconducting current loop. Red lines are radial trajectories of stellar flare protons approaching from the top. The mechanical protection plate in the middle of the figures (colored in blue) is a cylindrical water reservoir of 3 cm thickness. It absorbs

## Atmosphere

The temperature along the terminator is −10 to +25 centigrades. Liquid water exists in sufficient quantity to sustain life. The atmosphere is earthlike, with a protective ozone layer in the upper atmosphere. Close to the equator, strong jet streams transport heat from the day side to the night side, and water from melting glaciers and drifting oceanic ice flows toward the day side. At higher latitude, the weather is moderate, with winds and thunderstorms at the terminator. The incoming photonic flux density in the upper atmosphere is about 65% of that of the Sun on Earth, with an intensity maximum in the near infrared. Earth's daylight locus shown in Fig. 2.13 is shifted toward the right on Atlantis. At noon the sky appears light green to cyan, changing to yellow and orange at twilight. Scattering on clouds adds a touch of purple. The magnetic field of Atlantis is assumed to be about twice as strong as on Earth; it protects the atmosphere from severe mass loss during strong flares.

## Magnetic Shielding

No problems occurred during the EXO VR.2 missions thanks to the particle shields that work with a strong magnetic field created by a superconducting current loop. Proxima Centauri is known to undergo violent flares creating storms of charged particles, mostly protons. Values for the kinetic energy and the flux density of stellar outbursts were delivered by the *Astronomy* group. Figure 2.16a shows the field lines of current loops and the trajectories of incoming protons of 50 respectively 20 MeV kinetic energy, which are the most abundant in particle storms ejected from Proxima Centauri. The field deflects protons, creating a cylindrical safe space of about 20 m in diameter, two stories high. Even higher energy protons up to 200 MeV and alpha particles are well shielded, as well as particles approaching under a tilt angle of up to 20 degrees. The particles that could pass through the center of the coil are absorbed, ideally by a few centimeter thick layer of water. A mini device, only 30 cm in radius, is also available to protect individuals from sudden stellar flares, as sketched in Fig. 2.16b. The

---

**Fig. 2.16** (continued)  the central impinging particles. (a) 50 MeV protons deflected by a current loop of 3 m radius located 15 m above ground level (indicated as a disk/ring at $z = 0$). The nominal field at the origin is 2 Tesla, the safe space below the coil is a cylinder of ca. 20 m diameter, more than 6 m high. At ground level, the field is 28 mTesla. (b) Individual shield for short outdoor missions during flares, 10 MeV protons impinging. The nominal field at the origin is 6 Tesla. Right above the coils an aureole of glowing plasma appears, here only sketched in (b)

**Fig. 2.17** Protection coil with outward-pointing Lorentz force, creating tensile stress on the two indicated cross-sections. The red arrow is the direction of the magnetic field

energetic particles in rapid cyclotron orbits repelled at the center of the coil ionize air molecules, creating a gloriole sitting above the head of the shielded person. The 6 Tesla strong magnetic field exerts a radial (outward-pointing) Lorentz force on the current-carrying coils, resulting in a considerable tensile stress as indicated in Fig. 2.17. The tensile force on the indicated cross-sections is about $10^7$ Newton. A lightweight mechanical support—for the individual shields a 10 kg weighing Kevlar ring—protects the coils from rupture.

*Failure Analysis* The same principle should apply to the fuel harvesting of the Bussard ramjet. The request from *Construction* for coil parameters was first forwarded to *Astrophysics* to obtain the masses and charges of the interstellar gas, a prerequisite for the simulation. Unfortunately, we never received an answer from *Astrophysics*, and the request from *Construction* was not followed up. For the time being, it is not clear if there is a set of coil parameters for sufficient fuel scooping. If there is such a set, a ring of luminescent plasma should be visible at the entry port to the pipeline, close to the superconducting coil.

## Frontiers

*Failure Analysis* According to limited computing resources, the environment for EXO VR.2 was only implemented for a range of about 20 km around Cape Cod, the landing site. It was never expected that an expedition could reach the frontiers of the simulated environment, so the limiting coordinates were left visible on the circumference for further triangulation. In future implementations, it is planned to prevent the colonists from approaching the simulation limits, either by steep mountains or inaccessible canyons.

**Life Forms**

The EUROFORCE steering committee asked for a visual implementation of herbal life on the planet without giving further details on exobiological conditions. Whereas on earth, red light (about 12 photons) of solar radiation is necessary for the transformation of one molecule of carbon dioxide into oxygen, a different photosynthesis process utilizing a larger spectral range touching the near-infrared was programmed for Atlantis in order to compensate for the lower photonic input. The barren herbs and bushes modeled in EXO VR.2 are well adapted to the moderate climate at the terminator, much resembling vegetation at high latitudes on Earth.

***Failure Analysis*** Plants would probably not survive stellar flares. Therefore, the ESA inquiry panel asked for a more plausible implementation of plants on Atlantis, in order to satisfy the colonists. The rapid DNA repair mechanism of some extremophiles that support a radiation dose more than a thousand times higher than a human being could sustain should be implemented into lichen, grass, and grain. Some bushes could survive in valleys or cave entrances protected from particle storms where enough stray light from the sky prevails.

# Conclusions

The analysis has shown that the majority of a population is not able to detect slight deviations from physical reality. The theory determines what can be measured, as Einstein said, and most people have a fairly simple theory of their environment. The desire for an intact reality is so strong that even serious simulation errors are tolerated if they are not immediately evident. One example is the ad-hoc assumption of the development team that interstellar hydrogen can be collected in sufficient quantities with electromagnetic fields. The cover-up of the lack of knowledge under the pretext that the design principle is top-secret for military reasons has even dissuaded the physicist on board from further considerations. Clarke's principle that any sufficiently complicated technology is considered magic obviously applies to the natural sciences as well.

The crew was by and large fault-tolerant. But an inquisitive person can pass on a disturbing discovery to others and, so to speak, infect them, especially in an environment characterized by distrust and conspiracy theories.

The Ministry of Peace has ordered the continuation of the EXO VR.1 simulation on Atlantis at least until the lawsuit against EUROFORCE is settled. It is therefore of utmost importance to take every detail of reality into account in future versions of the EXO code. Modern quantum computers can fulfill this task. The real problems lie in the expertise of the analysts and in communication. Without prejudice to the court's decision, it is therefore recommended to intensify communication between and control of the programmers in order to avoid incoherence in the presumed reality, which can lead to mental disorder and irreversible social fracture of a forthcoming civilization.

## Appendix: Technical details

The equations used for the simulations described here, as well as more details and in-depth explanations, can be found at

    https://www.ustem.tuwien.ac.at/exodusincident

Printed in the United States
by Baker & Taylor Publisher Services